D0414712

eBooks
@Cardiff

INTEGRATING SEAPORTS AND TRADE CORRIDORS

Transport and Mobility Series

Series Editors: Professor Brian Graham, Professor of Human Geography, University of Ulster, UK and Richard Knowles, Professor of Transport Geography, University of Salford, UK, on behalf of the Royal Geographical Society (with the Institute of British Geographers) Transport Geography Research Group (TGRG).

The inception of this series marks a major resurgence of geographical research into transport and mobility. Reflecting the dynamic relationships between socio-spatial behaviour and change, it acts as a forum for cutting-edge research into transport and mobility, and for innovative and decisive debates on the formulation and repercussions of transport policy making.

Also in the series

International Business Travel in the Global Economy
Edited by Jonathan V. Beaverstock, Ben Derudder, James Faulconbridge and Frank Witlox
ISBN 978 0 7546 7942 4

Ports in Proximity
Competition and Coordination among Adjacent Seaports
Edited by Theo Notteboom, César Ducruet and Peter de Langen
ISBN 978 0 7546 7688 1

Railways, Urban Development and Town Planning in Britain: 1948–2008
Russell Haywood
ISBN 978 0 7546 7392 7

Transit Oriented Development
Making it Happen
Edited by Carey Curtis, John L. Renne and Luca Bertolini
ISBN 978 0 7546 7315 6

The City as a Terminal
The Urban Context of Logistics and Freight Transport
Markus Hesse
ISBN 978 0 7546 0913 1

For further information about this series, please visit www.ashgate.com

Integrating Seaports and Trade Corridors

Edited by
PETER HALL
Simon Fraser University, Canada
ROBERT J. McCALLA
Saint Mary's University, Canada
CLAUDE COMTOIS
Université de Montréal, Canada
and
BRIAN SLACK
Concordia University, Canada

ASHGATE

Published by
Ashgate Publishing Limited
Wey Court East
Union Road
Farnham
Surrey, GU9 7PT
England

Ashgate Publishing Company
Suite 420
101 Cherry Street
Burlington
VT 05401-4405
USA

www.ashgate.com

British Library Cataloguing in Publication Data
Integrating seaports and trade corridors. -- (Transport and mobility series)
1. Harbors--Economic aspects. 2. Shipping--Economic aspects. 3. Transportation corridors--Economic aspects.
I. Series II. Hall, Peter.
387.1'64-dc22

Library of Congress Cataloging-in-Publication Data
Integrating seaports and trade corridors / [edited] by Peter Hall ... [et al.].
p. cm.
Includes bibliographical references and index.
ISBN 978-1-4094-0400-2 (hbk) -- ISBN 978-1-4094-0401-9 (ebk) 1. Harbors. 2. Shipping. 3. Trade routes. 4. Infrastructure (Economics). 5. Business logistics. 6. Transportation--Planning. 7. Transportation geography. I. Hall, Peter Voss.
HE551.I48 2011
387.5'44--dc22

2010028233

ISBN 9781409404002 (hbk)
ISBN 9781409404019 (ebk)

Printed and bound in Great Britain by the MPG Books Group, UK

Contents

List of Figures

List of Tables

Notes on Contributors

Yann Alix is Director of IPER (Port Training and Research Institute), a component of the Normandy Business School in Le Havre where he also heads the Logistics Department. He is Associate Researcher for Logistel in Portugal and collaborates actively in different training and engineering projects in Africa for the Port Authority of Le Havre. After receiving his PhD from Concordia University and his Doctorate in Transport Geography from Caen University, he started a career as consultant for maritime innovation in Rimouski, leading applied research studies for Canadian and international port authorities, shipping companies and logistics entities. He is currently organizing professional port and logistic seminars in France and abroad in French, English, Spanish and Portuguese. He collaborates with the IMO to organize and host a five-week advanced port management seminar inviting 20 different countries each year. His research focuses on corporate management and strategies in the port and maritime industries as well as port-city relationships and freight inland corridors/transportation chains, especially on the African and South American continents.

Anthony Beresford was, prior to his current post at Cardiff Business School, a Research Associate at the University of Birmingham working on winter road maintenance, studying, in particular, the prediction of ice formation on roads. Anthony has traveled widely in an advisory capacity within the ports and transport fields in Europe, Africa, Asia and North America. He co-authored *Lloyd's Maritime Atlas of World Ports and Shipping Places* 15th and 16th Editions (1986 and 1989). He has been involved in a broad range of international transport-related research and advisory projects. In the UK, Anthony has assisted the DETR (subsequently the DfT) in, for example, the production of the discussion paper *Recent Developments and Prospects at UK Container Ports* published in July, 2001; he has acted as expert witness to the Public Enquiry into the proposed container terminal at Dibden Bay, Southampton, 2001-2005; and he has advised the Welsh Assembly on road transport and port policy options. Anthony has published over 150 academic papers and reports, with many of his papers appearing in front-rank academic journals. Within Cardiff Business School, Anthony is the Director for the highly successful MSc Schemes in International Transport, Logistics and Operations Management and Marine Policy.

Claude Comtois is Professor of Geography at the Université de Montréal, Canada. He has a degree in Political Science, a MSc in Geography from Laval University and a PhD from the University of Hong Kong for his research in the field of transportation. He is affiliated with the Research Centre on Enterprise Networks, Logistics and Transportation of the Université de Montréal. He has over 10 years experience as transport project director for the Canadian International Development Agency in China. Visiting professorships include more than 15 foreign universities. His teaching and research are centered on transport systems with an emphasis on shipping and ports. He has been involved in consultancy studies on marine policy, intermodal transport and environmental issues. He is a co-author of *Geography of Transport Systems* (Routledge) and has published many articles in scientific journals on port and marine transport. He currently supervises a project on sustainable port and inland waterways.

Jacques J. Charlier is a Research Associate with the Belgian National Fund for Scientific research and a Professor of Transport Geography at the University of Louvain-la-Neuve. In the last decade, he also spent more than three years as a Visiting, then Associate Professor at the University of Paris-Sorbonne. He has developed a strong research interest in ports in general and in port hinterlands more especially, in developed as well as in developing countries. His PhD dealt with the hinterlands of French seaports, and he has authored more than 100 academic papers in the field of port geography in his 35-year-long career.

Gustaaf De Monie graduated in 1965 from the University Antwerp with a degree in Applied Economics and started his professional career with a leading Antwerp Terminal Operator. In 1970 he joined the Ports Section of UNCTAD where he was responsible for research, training and project development. In 1989 he established himself as an independent consultant. He is presently acting as an analyst and international adviser on port and shipping matters and works for international organizations such as the World Bank and IFC, global terminal operators and port authorities. He is Programme Advisor to the Le Havre Port Training Institute and Course Coordinator for 'Terminal Management' in the ITMMA Master of Science programme of Antwerp University. Last but not least, he is Project Director for 'Policy Research Corporation', an Antwerp-based economic think-tank specializing in transport issues. Dr De Monie's key qualifications include a broad experience in project management, port operations, port reforms and performance measurement. He is the author of numerous research studies and papers on port management, operations and container terminal development. He is an organizer of port training programmes and team leader for the implementation of port development and re-structuring projects.

Antoine Frémont, PhD, is holder of the Agregation of Geography, Doctor of Geography (PhD from Le Havre University in 1996) and has accreditation to supervise research (University of Paris Pantheon-Sorbonne in 2005). He works as a Director of Research at the INRETS (The French National Institute for Transport and Safety Research). He specializes in maritime transport issues and focuses more specifically on shipping lines strategies, maritime networks, hinterlands and the organization of combined transport chains and the development of logistics within the maritime sector. He is currently managing a research program about container barges and the hinterland of French ports. He is also involved in an OECD workshop about the reliability and levels of service of the surface transport networks. He has published several books, chapters of books and academic papers. He participates regularly in international conferences.

Elisabeth Gouvernal is Senior Researcher in the French National Institute on Transport and Security, and head of the research unit 'Production Systems, Freight and Logistics'. She holds a doctorate in economics and planning. She has held positions in the World Bank and the European Commission on the competitiveness of transport chains, and in the shipping company, CMA-CGM. Her current research focuses on maritime transport, the organization of transport chains, multimodality and problems of inland haulage to and from ports. Her present research focuses on port governance and devolution, shortsea shipping and railway shortlines. She lectures at the University of Paris I and the Ecole Nationale des Ponts et Chaussées on maritime transport and ports.

Emmanuel Guy is Professor in the Department of Management Sciences of the University of Quebec at Rimouski, where he holds the Research Chair in Maritime Transportation. In this position, he works in association with stakeholders of St. Lawrence's maritime industry, researching the development of effective shipping policies. Initially trained as a deck officer, he served at sea for three years: sailing on square-rigged training vessels, bulkers and a survey ship. His academic formation includes a degree in Anthropology from the Université de Montréal; a masters in Maritime Resources Management from the University of Quebec at Rimouski and a doctorate in Transport Geography from the University of Montreal. He also conducted post-doctoral work at his current institution looking at environmental issues in Canadian maritime transportation. His work has appeared in *Maritime Policy and Management*, *Maritime Economics and Logistics*, *Journal of Transport Geography*, *Journal of Ocean Technology* and *Organisations et Territoires*.

Peter Hall is Professor in the Urban Studies Program at Simon Fraser University in Vancouver, Canada. Before completing his doctoral studies in 2002 in city and regional planning at the University of California, Berkeley, he worked in the Economic Development Unit of the Durban Metropolitan Council. His current research examines the connections between shipping and logistics chains, transport sector employment and the development of port cities. His research on ports has been published in several academic journals, including *Regional Studies*, *GeoJournal*, *Journal of Urban Technology*, *Environment and Planning A*, *Maritime Policy and Management*, *Economic Development Quarterly*, *Tijdschrift voor Economische en Sociale Geografie*, and *Economic Geography*, as well as in edited volumes.

Trevor D. Heaver is Professor Emeritus, University of British Columbia, where he was Director at the Centre for Transportation Studies. He is a Visiting Professor at the University of Antwerp and lectures frequently internationally. He is a founding member and a Past-President of the International Association of Maritime Economists and a Past-Chairman of the World Conference on Transport Research. He has published widely on transportation, logistics and transportation policy. Dr Heaver's current research interests focus on issues related to liner shipping, international logistics and intermodal transport. He has served and still serves as an advisor and consultant to public and private organizations in Canada and internationally.

Michael C. Ircha is Professor and Associate Vice-President Emeritus at the University of New Brunswick, Adjunct Research Professor at Carleton University and Senior Adviser to the Association of Canadian Port Authorities. In the past, Michael served as planning director, city engineer and chief administrative officer in municipalities in Nova Scotia and Ontario. He holds degrees in Civil Engineering, Urban Planning, and Public Administration from Queen's University; a certificate in international strategic studies from Canada's National Defence College; and a doctorate in Ports Administration and Planning from the University of Wales, Cardiff. Dr Ircha is the 2004 recipient of the New Brunswick Lieutenant-Governor's Award for Excellence in Public Administration. A frequent international lecturer, he was named "Distinguished Transportation Professor" by the International Intermodal Expo, and has received several research awards for his transportation publications. On an annual basis, he provides port management courses at the UN's World Maritime University and Shanghai Maritime University. Dr Ircha is a fellow of the Engineering Institute of Canada and a Fellow and Past-President of the Canadian Society for Civil Engineering. Mike offers consulting services in policy and planning issues related to marine transport and ports, as well as municipal administration and urban planning.

Wouter Jacobs received his MA in Spatial Planning at the Radboud University Nijmegen in 2001. He continued his research at the same university and successfully defended his PhD-thesis entitled *Political Economy of Port Competition: Institutional Analyses of Rotterdam, Southern California and Dubai* in 2007. Jacobs currently works at the Utrecht University as an Assistant Professor. His current research focus is on the relationship between port clusters and advanced producer services in international value chains.

Hans Koster obtained his Bachelors degree in Economics and Business at the Erasmus University Rotterdam in 2008. In 2009 Hans obtained his Masters degree in Economics, cum laude. In this year he started a PhD study at the VU University which focuses on the nature of transport corridor development in the Netherlands. His fields of interest are urban economics, network analysis and port economics and geography. He wrote his bachelors thesis about commodity specialization in ports which led to a publication in *Regional Studies*. For two years, he has cooperated with Wouter Jacobs on a project which investigated the relations between port clusters and maritime advanced producer services.

Frédéric Lapointe is a full-time research agent at the Research Chair in Maritime Transportation at the University of Quebec at Rimouski. He holds a BSc in Geography with a specialization in planning. He has also completed a Masters degree in integrated management of maritime resources – concentration in shipping – at the Université du Québec à Rimouski. His research essay analyzes the application of data management systems to marine transportation. His work with the Chair consists of public policy analysis, with a strong focus on trade corridors and gateways, short sea shipping and marine transportation statistics. More generally, his research fields focus on maritime transport, transport and territorial development.

Valérie Lavaud-Letilleul, a former student at the Ecole Normale Supérieure (Ulm, Paris), is holder of the Agregation of Geography and a doctorate of Geography (PhD from University of Paris Pantheon-Sorbonne in 2002). She is Associate Professor at the University of Montpellier 3. She works in association with the INRETS (The French National Institute for Transport and Safety Research). She specializes in port and city-port development issues and focuses more specifically on port and non-port actors' strategies in city-ports, new scales in port planning, the integration of port activities within coastal areas and more generally speaking, in coastal and marine management. She is currently involved in a research program about the evolution of port governance in French, Italian and Spanish ports. She is also involved in a program about the impact of environmental issues in port development. She has published 10 chapters in books and eight papers in academic or professional reviews.

Robert J. McCalla is Professor of Geography at Saint Mary's University in Halifax, Canada and Adjunct Professor in the Marine Affairs Program at Dalhousie University. His research and teaching interests are in maritime transportation. He has published in journals such as *Journal of Transport Geography*, *Canadian Geographer*, *Maritime Policy and Management*, *Geoforum*, and *Tijdscrift voor Economische en Sociale Geografie*. He is the author of *Water Transportation in Canada* (Formac 1994). He served for four years as an External Examiner of the World Maritime University (Malmo, Sweden) and Shanghai Maritime University. Dr McCalla was born in Windsor, Ontario, Canada. His BA degree is from the University of Western Ontario; his PhD is from the University of Hull, England where he was a Commonwealth Scholar. He is married with two married sons.

Theo Notteboom is President of ITMMA (Institute of Transport and Maritime Management Antwerp, an institute of the University of Antwerp, Belgium).

He is also affiliated with the Faculty of Applied Economics of the University of Antwerp and is part-time Professor in Maritime Transport at the Antwerp Maritime Academy. In 2007, he held the MPA Visiting Professorship at Nanyang Technological University in Singapore and he still holds a Visiting Professorship at Dalian Maritime University in China. He has published widely in academic journals and books on transport and maritime economics, transport geography and transport policy, including market organization, spatial developments, maritime transport and inland transportation. His research output covers applications in Europe, Asia/China and North America. He is a member of the editorial boards of *Journal of Transport Geography*, *Maritime Policy and Management*, *Maritime Economics and Logistics* and *WMU Journal of Maritime Affairs*. He is Council member of the International Association of Maritime Economists (IAME) since 2006, Chairman of the Board of the Belgian Institute of Transport Organizers (BITO), Chairman of the steering committee of Asian Logistics Round Table (ALRT) and a member of the Board of BIVEC (Benelux Inter-University Group of Transport Economists).

Athanasios A. (Thanos) Pallis is Assistant Professor at the Department of Shipping, Trade and Transport, of the University of the Aegean, Greece. He is responsible for academic courses on Port Management, Economics and Policy, and the holder of a Jean Monnet grant on European Port Policy. He also holds a Visiting Professorship at the Centre for International Trade and Transportation, Dalhousie University, Halifax, Canada, and lectures at the University of Antwerp, ITMMA. Since 2008 he is a Fulbright Scholar at the Centre for Energy, Marine Transportation and Public Policy (CEMTPP), Columbia University, New York. In 2008 he co-authored a study focusing on the concessions of port terminals that won Palgrave Best Paper in *Maritime Economics and Logistics*. His book publications include *Maritime Transport: The Greek Paradigm*, *European Port Policy* (Edward Elgar; published in English, Japanese and Greek), and *The Common EU Maritime Transport Policy* (Ashgate 2002). He is a regular contributor to European Seaport Organisation (ESPO) and OECD discussions on the future of seaports, the Secretary General of the Hellenic Association of Maritime Economist (HAME) and a Council member of the International Association of Maritime Economists (IAME).

Francesco Parola started his studies in Maritime and Transport Economics at the University of Genoa, Italy, where he got his Bachelor degree in early 2001. During his PhD studies he was a Visiting Researcher at the Center for Maritime Economics and Logistics of the Erasmus University in Rotterdam. He defended a doctoral thesis (2005) about the strategies of transnational container terminal operators in the port reform era. His favourite research topics are liner shipping, maritime logistics and port governance. Currently he is Researcher at the 'Parthenope' University in Naples, lecturer at the Faculty of Economics in Genoa, contract Professor at the Italian Naval Academy in Livorno, and member of the Italian Centre of Excellence for Integrated Logistics. He is author and co-author of many papers and articles in conference proceedings and peer-reviewed international journals.

Jean-François Pelletier works at the Normandy Business School in Le Havre where he teaches Logistics and Maritime Transportation Management. He holds a MSc degree in Management of Maritime Resources from the Université du Québec à Rimouski. He is also working on his doctoral thesis within the École Doctorale Ville et Environnement in Paris and is affiliated with the Systèmes Productifs, Logistique, Organisation des Transports et Travail laboratory of the Institut National de Recherche sur les Transports et leur Sécurité (INRETS). His research interests concern freight transportation analysis.

Stephen Pettit graduated with a BSc Honours degree in Maritime Geography from Cardiff University in 1989 and in 1993 he was awarded a PhD from the University of Wales. In the last decade he has been involved in a range of transport related research projects. Notably, he was involved in a groundbreaking project for the Department of Transport, studying the UK economy's requirements for people with seafaring experience. He has been involved in a wide range of international transport related projects for EU DGVII (DGTREN). His broad research interests include port development and port policy and the logistics of humanitarian aid delivery.

Jean-Paul Rodrigue is an Associate Professor in the Department of Economics and Geography at Hofstra University, New York. His research interests primarily cover issues related to freight transportation, logistics and globalization, particularly over North America and Pacific Asia. His recent work is focusing on the integration of maritime and inland freight distribution systems through the setting of gateways and corridors and how containerization has changed freight distribution. Dr Rodrigue has developed a high impact transport website (*The Geography of Transport Systems*) and was chair of the Transport Geography Specialty Group of the American Association of Geographers (2004-2006). He is also on the international editorial board of the *Journal of Transport Geography*, the *Journal of Transport and Land Use* and acts as the Van Horne Researcher in Transportation and Logistics.

Brian Slack is a graduate of the London School of Economics and McGill University. His academic career has been anchored at Concordia University, Montreal. A transport geographer with an interest in container shipping and intermodal transportation, he has published widely on these topics and has occupied visiting academic positions around the world. His present research focuses on three areas: port governance and devolution, shipping and sustainability, and the cultural dimensions of container shipping.

Thomas K. Vitsounis is a PENED Research Fellow at the Department of Shipping, Trade and Transport, of the University of the Aegean, Greece. He participates in the research programme 'Strategies and Competitiveness of Greek Shipping and Greek Ports' while he works on a PhD thesis examining port service providers and users relations. He also contributes to the teaching activities of the undergraduate and postgraduate programmes of the Department. His scientific interests include port economics, policy and management, ports competition and competitiveness, and especially port performance measurement and port users' satisfaction and value. Thomas is the author of scientific papers that examine these issues, as well as coastal shipping. He has also participated in international and national research programmes that examine the structure of port, coastal and maritime markets and the development and evaluation of appropriate national and European policies. He is also a member of the International Association of Maritime Economists (IAME), and acts as the Secretariat for the Hellenic Association of Maritime Economist (HAME).

James J. Wang is Associate Professor at the Department of Geography, the University of Hong Kong. Born in Beijing, he received his Bachelor in Economics from the People's University of China, MPhil from the University of Hong Kong, and PhD from University of Toronto. Currently he is council member of Hong Kong Society for Transport Studies, and Fellow of Chartered Institute of Logistics and Transport (FCILTHK). His research area is transport geography, with special interests in port development and port-city relations. He has published widely in many internationally refereed journals and is editorial board member of *Journal of Transport Geography* and *Transportmatrica.* He is chief editor of *Ports, Cities, and Global Supply Chains* published by Ashgate in 2007. As a port-city specialist, Dr Wang has participated in port-city planning projects and strategic studies for more than 25 Chinese and other Asian port cities and regions.

Su-Han Woo graduated with a BA degree in Economics from the Seoul National University, Republic of Korea in 1998, and is currently a PhD Researcher in Cardiff Business School. He also has been working in ministries dealing with transport in the Korean Central Government as a Deputy Director since 1999, and has been involved in a wide range of maritime transport related policies. Notably he was deeply involved in deregulation and privatization in Korean ports including port labour reform. His broad research interest is international transport and logistics, focusing on ports in supply chains and the operation and management of those chains.

Chapter 1
Introduction

Peter Hall, Robert J. McCalla, Claude Comtois and Brian Slack

The book focuses on the theory and practice of integrating seaport gateways and associated trade corridors into global value chains, taking into account the current world economic condition. It addresses a key question that confronts the shipping and port industries as well as public authorities: how can the benefits of maritime trade to the companies and institutions directly involved in trade, as well as the port city-regions where the transfers take place, be increased? What strategies and developments are these actors pursuing as they seek to integrate seaports and trade corridors, and what are the likely consequences of their actions? This question is being posed following a global economic recession and trade downturn, and in the context of contemporary policy frameworks of developing trade gateways and corridors whose goals are to generate economic benefits and efficiencies rather than simply to maximize traffic volumes.

Seaport gateways and the corridors which connect them to widely dispersed hinterlands are of vital and essential importance to international trade and the world economy. No economy can survive without trade, and investment in fixed infrastructure features prominently in the economic recovery plans of countries around the world. At the same time, there will surely be a huge opportunity cost to be paid for investing in the wrong trade gateway or corridor, or for investing in the wrong way, especially if this locks in a sub-optimal or dependent path of development.

Maintaining and increasing trade involves maritime transport – that is, the trade is not just land based. It is generally accepted that more than 90 per cent of world trade is handled in ships. In 2007 over 7.4 billion tons of cargo were carried in ships around the world – all of which had to be handled in seaports as either loaded or unloaded commodities. Distributing these goods to ultimate land destinations or bringing the goods to seaports from inland origins is organizationally complex involving multiple actors. It is important to understand how this movement is organized, and what role shipping lines, ports and corridor-actors play in this regard.

In the new economic reality all aspects of global supply chains need to be re-examined. Trade is down, ships have been laid up or sail light, ports have decreased throughput, inland trucking and railroads are suffering, and the practice of adding value to trade goods and services associated with those goods is marginalized. How can seaports acting as gateways to trade and the associated inland infrastructure and services in the corridors connecting gateways to origins and destinations react to the changing environment? This book attempts to put into perspective the opportunities and challenges facing seaport gateways and corridors now and in the future.

Themes

Integration is a complex term having different meanings to different users (see Chapter 13). For planners and public administrators, integration has the connotation of greater coherence and consistency across policy domains such as land use, public works, environmental regulation and public finance; for economists, vertical integration along value chains is often contrasted with horizontal integration across a single sector; and for scholars of globalization, integration refers to the blurring of national boundaries that have accompanied greater flows of people, goods, finance and ideas. In *Integrating Seaports and Trade Corridors*, the contributors have focused on four dimensions of integration which provide the themes of the book, namely the new economic context, value, physical connectedness, and collaboration and coordination. Each chapter picks up on one or more of them.

The first theme of the book is the *new economic context* for ports and corridors. If economic integration was a hallmark of globalization, and if coupling and fallout rather than decoupling and insulation were characteristics of the global economic crisis, then we cannot understand the future operating context of ports and corridors without paying attention to the new global economic context. This new context for world trade, shipping and ports is one of economic uncertainty. It cannot be assumed that the rapid growth in world trade that preceded the crisis will resume undiminished, nor that the patterns of trade growth in the future replicate those of the past (see Chapter 3). Those who plan and operate ports and corridors will need to pay attention to both existing and emerging patterns of trade, and most especially to the strategies of the most powerful actors. How we got to a situation where the fortunes of ports were so deeply embedded in the global economy, both directly in terms of throughput, and indirectly via the financial sector, is the focus of Chapter 2. Each of the other chapters in the book explicitly or implicitly considers how seaports and corridors can achieve effective and beneficial integration in a most challenging economic context.

Second, in uncertain economic times, communities, governments, shareholders and other actors are compelled to ask hard questions about the *value* of seaports and trade corridors, especially since trade activity declined even more rapidly than overall economic activity in the global recession. As these stakeholders consider where to invest in future, they seek a more complex and nuanced understanding of the value of port and corridor infrastructure. The value theme features prominently in Part II, which addresses how changes in the organization of intermodal transportation call for re-thinking the measurement of performance and value-added in logistics chains (see Chapters 8 and 9). Chapters explore the "what" and "why" of port performance measurement, in addition to the "how-to" questions. Discussion of the notion of value is also found in chapters dealing with the relationship between transshipment and added value logistics (Chapter 5) and the geography of advanced maritime services (Chapter 6), as well as in the case studies in Part III of the book.

The third theme of the book addresses a traditional concern of port-hinterland relationships: that of *physical connectedness*. Questions of hinterland access and corridor development gained prominence before the recent global recession in the context of rapid trade expansion. There is, however, no reason for thinking that the resultant processes of shipping lines securing preferential access to gateways (see Chapter 4), as well as port regionalization and hinterland expansion (see Chapter 15), will be simply reversed in the face of a decline in trade volumes. Arguably, the importance of physical connectedness may increase, as governments make infrastructure spending available as part of their economic stimulus plans, as the return of high energy prices encourages a switch away from road-based transportation, or as ports look to elaborate their corridor connections in the newly competitive environment. Several of the international case studies in Part III identify these trends. One especially intriguing notion is the potential emergence of new connections beyond the gateway-corridor scale – for example, the potential for overlapping corridors and webs of connections at a continental scale (see Chapters 11 and 14). These observations resonate strongly with the notion of the multi-jurisdictional mega-region that has been gaining currency both in Europe and in the United States.

The fourth theme addressed in this volume concerns *collaboration and co-ordination*, or what can be understood as the inter-organizational analogue of physical connectedness. Global supply chains imply, by definition, more complex relationships between shippers, carriers and public authorities. As port users respond to global economic conditions through new plans, policies and investments, they must collaborate with a wider array of actors for effective and timely implementation. They must also act to ensure the provision, management and sharing of timely and accurate information. Several chapters explore whether there is a renewed role for port authorities, governments and other actors in this regard (see Chapters 7 and 10). These and other chapters also explore the appropriate level of collaboration, and what are the appropriate mechanisms for coordination, in the current context (see Chapters 12 and 16).

Organization of the Book

The book is divided into three parts. The first part focuses on *global economic change*, including the sources and implications of current economic conditions, as well as other longer term changes in the global economy and the maritime shipping industry for ports, corridors and value chains. The second part turns to the theoretical and practical dimensions of *measuring and improving gateway and corridor performance*, relating these measures not only to throughput but also to enhanced services among shipping lines, terminals and inland carriers. The third part presents *international case studies* of seaport gateways and corridors, three focused on Canada, one each on Africa, Europe and China.

Part I opens with a pair of chapters that respectively look backward and forward from the recent global economic crisis. In Chapter 2, *Economic Cycles in Maritime Shipping and Ports: The Path to the Crisis of 2008*, De Monie, Rodrigue and Notteboom trace out the perverse integration of ports into two aspects of global economy. First, they show how global financial and trade imbalances contributed to a boom, and arguably to overexpansion in shipping and port capacity. Second, they trace the increasing financialization of the ports sector whereby terminals and other transportation facilities increasingly became regarded as tradable financial assets. Connecting these two "perverse" integrations was the tendency of forecasters to produce ever-inflated projections of future growth in container throughput. When the global financial bubble burst, as it had to, the ports sector was not far behind, with falling throughput and falling asset prices the outcome.

In the following chapter, Notteboom, Rodrigue and De Monie look forward, exploring *Organizational and Geographical Ramifications of the 2008-09 Financial Crisis on the Maritime Shipping and Port Industries*. The authors take on the important question of whether the recession of 2008-2009 will mark a turning point in global distribution systems and value chains, or come to be understood as a temporary correction in trends that resume unabated. They come down in favour of the former, paradigm-shifting, perspective while remaining conscious of the inherent dangers in any act of prediction. Whatever the outcomes, they are insistent that key actors such as shipping lines and terminal operators – and the strategic choices they take with respect to investment, integration, routing and infrastructure – will play a key role in the emerging patterns of concentration and dispersal.

Frémont and Parola's chapter on *Carriers' Role in Opening Gateways: Experiences from Major Port Regions* provides recent historical support for this assertion, and is a timely reminder of the active and strategic role of large maritime carriers in the process of gateway formation. They analyze container throughput data for major lines and alliances from 1994, 2002 and 2006 using the Gini coefficient to trace how leading lines both concentrate and deconcentrate their operations within different port ranges. Their findings highlight that all shipping lines seek to secure dedicated gateways within different port ranges, but that there is diversity within the strategies employed to achieve this goal. The findings also hold important implications for how particular firms and the ports they use may emerge from the global recession; in particular, while vertically integrated lines with significant terminal operations have a scale advantage, they may lack the flexibility demanded by the new context.

The fifth and sixth chapters raise similarly hard questions about the possible trajectories of economic recovery and development for ports, and their host city-regions. Gouvernal, Lavaud-Letilleul and Slack critically analyze the claim that transshipment hubs – themselves a consequence of lines routing strategies – will develop as growth poles, attracting logistics activities to promote economic development. In *Transport and Logistics Hubs: Separating Fact from Fiction*, they argue that a proper appreciation of the distinction between the goals of carriers who use hubs to maximizing economies of scale in shipping, and those of producers and distributors who use logistics hubs to optimize production chains raises challenging questions about the possibilities of promoting logistics and economic development in transshipment hub ports.

In Chapter 6, *Port, Corridor, Gateway and Chain: Exploring the Geography of Advanced Maritime Producer Services*, Hall, Jacobs and Koster explore the relationship between advanced maritime services and the geographic entities that demarcate port systems. Turning from the traditional focus on employment in the goods-handling transport sub-sectors, they ask whether advanced transportation services firms such as ship finance, insurance and maritime law, and logistics consulting also concentrate within gateway cities. While not answering this question conclusively, their analysis of data on the network connections of advanced maritime service firms, employment in maritime services in the United States, and case study material for Vancouver, Canada, suggests that there is no necessary spatial relationship between port throughput and advanced maritime services. As with Chapter 5, this conclusion is a cautionary note for policy-makers about the assumed local economic development benefits of port activity.

Part II contains four chapters that wrestle with the nuts and bolts of *measuring and improving gateway and corridor performance*. However, each also takes seriously the theoretical conundrum that debilitates so much of the existing literature on port performance, namely, how to conceptualize, measure and improve performance of port facilities that are themselves inextricably connected to wider systems over which they have little or no control. In making concrete recommendations for action, each of the contributors to this section call for increasing the visibility of the various elements of the value chains that traverse ports and trade corridors today.

In Chapter 7, Comtois and Slack draw lessons from the waterfront for *Measuring Port Performance*. After reviewing the academic literature on port performance measurement, and considering the experiences of port authorities and government agencies that have used key port performance indicators (KPPI), they conclude that the process of developing indicators is at least as important as the technical and methodological task of creating them. Hence, they apply considerations of feasibility, robustness, usability, transparency and readability to their assessment of a system of key performance utilization (KPUI) indicators proposed for Canada's bulk ports. Another important contribution of this chapter is that it addresses bulk shipping, a sector that is often overlooked in favour of containerized goods.

Containers, and the various models of terminal governance that have emerged under containerization, are the central focus of Pallis and Vitsounis's chapter on *Key Interactions and Value Drivers Towards Port Users' Satisfaction*. Proceeding from the observation that (container) ports are now populated by diverse actors organized in a variety of contractual and operational relationships, they ask what underlying interactions and value drivers create satisfaction for each of the users in this complex system. In so doing, they reject the notion that a singular measure can provide sufficient information to improve port performance. Using the case of four international European ports (Antwerp, Piraeus, Thessaloniki and Zeebrugge), they argue that a Business Performance Management framework can provide a structure to assess user value. Another noteworthy finding of this chapter that resonates with the argument of Heaver (Chapter 10) is that there is a mismatch between ports users who greatly value information management systems, and port authorities – at least those surveyed – who do not see information management as a core element of this business and operational activities.

In the era of global supply chains, ports are confronted with a new and diverse set of challenges, and in response they have adopted a new set of strategies, including the provision of value-added and intermodal services and interaction with other supply chain members designed to integrate ports more firmly within those supply chains. Is it worth the effort, and how can practices be monitored and improved? In Chapter 9, Beresford, Woo and Pettit take on the complex task of establishing whether actions by ports to integrate themselves into supply chains actually improve port performance. They make use of structural equation modeling to test a conceptual model on the relationship between the integration of ports into value chains, and conclude that ports are positively integrating into supply chains in order to acquire competitive advantage and perhaps to diversify their portfolio. At the same time, they echo the findings of Pallis and Vitsounis (Chapter 8), showing that the integration of ports into supply chains takes place to different degrees and in different ways, driven by a variety of commercial and other motives.

In Chapter 10, Trevor D. Heaver turns his attention to the question of *Coordination in Multi-Actor Logistics Operations: Challenges at the Port Interface.* Heaver focuses on the new roles of an old type of organization, the landlord public port authority. Using the case of Port Metro Vancouver, he argues that port authorities have always played a role in coordinating the activities of the members of their port community. In the new ports' context, this implies that they are required to extend their involvement in operations beyond infrastructure provision at or near the waterfront, to provide the information that will allow users to coordinate activities in more complex and extensive activity systems. At the same time, however, Heaver also notes the important role of shippers in value chains, and hence the need to engage them in multi-actor collaborations.

Part III presents *international case studies* from four continents. The cases can be read in several ways. First, each of the cases picks up two or more themes of the book. The theme of the *new economic context* informs all the cases; readers especially interested in the *value* theme may want to start with Chapters 11 and 12; readers especially interested in *collaboration and coordination* may want to start with Chapters 13 and 14; and readers especially interested in *physical connectedness* may want to start with Chapters 15 and 16.

Second, by telling the story of an actual place, each case study elaborates the idea that while seaport gateways have much to gain from their relationship to trade corridors, integration is neither automatic, nor is it necessarily a smooth or uncontested process. Rather, integration requires hard work by committed and powerful actors.

And third, since three of the case studies address the Canadian experience with, and debates about, integrating seaports and trade corridors, they (Chapters 12, 13 and 14, along with Chapters 6, 7 and 10) together create a meta-case that can be contrasted with experiences elsewhere. In particular, the Canadian cases highlight the challenges that arise in a complex and fragmented governance system, where port authorities are federal not-for-profit corporations but where many transportation-related powers and responsibilities lie with the provincial authorities and where local industry groups have assumed an increasingly vocal role in transportation planning. Yet, as the African, Chinese and European cases also illustrate, the challenges of gateway and corridor governance are no less important elsewhere.

In Chapter 11, *Benchmarking the Integration of African Corridors in International Value Networks*, Pelletier and Alix ask whether a gateway/corridor with Libyan backing can provide a superior service for resource-exporters based in the sub-Saharan nations of Chad and Niger. The Libyan effort finds itself in competition with gateway/corridors originating to the south (Cotonou, Lagos and Douala) and east (Port Sudan) of these land-locked nations. Using a series of operational and cost benchmarks, they compare the alternative routes. At the same time, they broaden our understanding of value in trade corridors by including assessments of stability, safety and environmental conditions which are identified as critical factors for some users. Another contribution of the chapter is that it enriches our understanding of the relative importance of overland transportation and maritime transport in the competitiveness of African value chains.

In the first chapter focused on the recent Canadian experience, Guy and Lapointe consider what engages different stakeholders in policy processes designed to improve gateway and corridor performance. In *Building Value into Transport Chains: The Challenges of Multi-Goal Policies*, they explore the case of gateway initiatives within the St. Lawrence Great Lakes system. Their assessment is that, to date, participants in these initiatives have not yet moved beyond their original, individualistic perspectives on the transportation challenges facing the region. Instead, these perspectives are deeply held as evidenced by their persistence despite the economic downturn. Guy and Lapointe argue that in order for a gateway and corridor initiative to really succeed, it must be valued by participants as something more than the sum of its individual parts.

McCalla's chapter on *Perspectives on Integrated Container Transport: The Canadian Example* suggests that such clarity and transparency in policy-making is unlikely to happen easily. His assessment of the evidence and findings of the recent review of containerized freight transport undertaken by the Canadian Senate

Standing Committee on Transportation and Communication finds that participants have very different understandings of the concept of "integration". He identifies five uses of the term integration, related respectively to transport planning and policy, supply/logistics chains, intermodalism, social and environmental concerns and economic development. While some categories of participant favoured particular understandings of integration, no single perspective dominated. One implication of this observation is that it will be difficult to create structures to promote the integration of gateways and corridors without agreement on basic questions such as who or what do we wish to integrate, and how does integration differ from other forms of coordination and collaboration?

In the third Canada-focused case study, *Trade Corridors and Gateways: An Evolving National Transportation Plan*, Ircha provides a more optimistic vision of gateway and corridor integration based on a bottom-up approach to devising a national transportation plan. While recognizing the inter-governmental challenges that have for long bedeviled attempts to resolve such issues as differing national and provincial truck weights and dimensions, taxes for road and rail, and intermodal integration, Ircha sees in the various regional gateway and corridor initiatives (the Asia-Pacific Gateway and Corridor Initiative, the Ontario-Quebec Continental Corridor and Gateway and the Atlantic Gateway) the emergence of a hierarchical structure of local, regional and national gateway councils comprised of private transportation providers, users and others. An especially intriguing aspect of his observation is the fact that the three major corridors overlap in such as way that they cover all of Canada's southern and most populous regions.

In Chapter 15, Charlier turns our attention to the relationship between policy-making and the physical connectedness between port and hinterland. In *Hinterlands, Port Regionalisation and Extended Gateways: The Case of Belgium and Northern France*, he traces both the theoretical and policy origins of the concept of the extended gateway. These systems of connected inland logistics centres have been used by the ports of Antwerp, Zeebrugge and Rotterdam to anchor themselves to their Belgian, Dutch and German hinterlands. Charlier then traces how these same ports are using the extended gateway concept to anchor themselves to the French portions of their hinterland. In response, French ports such as Dunkirk and Le Havre have begun to develop their own extended gateways oriented towards the Paris region. As such, extended gateways are more than just systems of physical connection between port and hinterland; they are also a strategic arena in which ports, and their users, compete with each other.

The volume ends with Wang's case study of *Entrepreneurial Region and Gateway-making in China: A Case Study of Guangxi*. Wang draws our attention away from the iconic Chinese port-regions (such as Hong Kong-Shenzhen and Shanghai-Ningbo) to those places that have yet to emerge as centres of economic globalization. His case study of Guangxi illustrates the active role of the regional state in gateway-making and promotion. That post-reform governments in China are able to direct considerable resources to such projects is clear, but Wang warns us that the effectiveness of these practices have yet to be proven.

And Finally … a Note on Canadian Content

In closing, we want to highlight the Canadian content of this volume. When Canadians speak about their country, both its maritime frontiers and its vast land mass feature prominently. The national motto, *A Mari usque ad Mare* ("From Sea to Sea"), recognizes that the inclusion of Canada's Pacific-facing Province of British Columbia in the confederation was only made possible, in part, by a promise that its ports would be integrated into the national railway system that already extended from the Atlantic coast through central Canada. Today, it is increasingly common to hear the phrase "from sea to sea to sea" which adds recognition of Canada's arctic coastline and peoples – not to mention the prospect for arctic shipping and all the benefits and risks that this implies.

The challenge of connecting these seaport gateways to the land that lies between is an equally important Canadian theme, for in the words of former Prime Minister William Lyon Mackenzie King, "if some countries have too much history, we have too much geography". Most recently, local, provincial and national interests have converged on the notion of gateways and corridors as organizing principles for transportation infrastructure spending for economic growth. As noted, several chapters in this book have explored the ongoing evolution of this policy, its contested meanings, its potential impacts and its limitations.

Today, the small, open Canadian economy is heavily dependent on land-based trade with its southern neighbour. However, in this context, international marine trade is especially important for economic diversification, with the marine sector accounting in 2007 for almost half of the value of all merchandise trade with countries besides the United States (Transport Canada 2009a). Canada's ports handle a wide variety of bulk cargoes, with container operations concentrated in Vancouver, Montréal and Halifax.

It is thus appropriate that Canadian content holds an important, but hopefully not overwhelming, presence in this book. Of course, it is also no accident. The editors are located in the three major Canadian port regions, from Halifax in the east to Vancouver in the west, and in Montréal in central Canada. The chapters in this volume were first presented at a conference in Montréal, 10-12 June 2009. The conference and book project would not have been possible without the generous funding support of Transport Canada and additional support from the Interuniversity Research Centre on Enterprise Networks, Logistics and Transportation (CIRRELT), Université de Montréal, and the Social Sciences and Humanities Research Council of Canada.

Finally, we would to thank Ashgate and their *Transport and Mobility Series* for their support of this book, and scholarship on ports and maritime transport more generally. We also wish to acknowledge the assistance of Carolyn Ruhland in formatting the text and preparing of the index. This volume builds upon the ideas presented at earlier conferences in Le Havre, Hong Kong and Antwerp-Rotterdam, and in the publications that followed them: two special issues of *Les Cahiers Scientifiques du Transport* (2004) and *Belgeo* (2004), and two volumes in this series, *Ports, Cities and Global Supply Chains*, (2007) edited by J. Wang, D. Olivier, T. Notteboom and B. Slack, and *Ports in Proximity: Essays on Competition and Coordination Among Adjacent Seaports*, (2009) edited by T. Notteboom, C. Ducruet, and P. de Langen.

PART I
Global Economic Change: Implications for Ports, Corridors and Value Chains

Chapter 2

Economic Cycles in Maritime Shipping and Ports: The Path to the Crisis of 2008

Gustaaf De Monie, Jean-Paul Rodrigue and Theo Notteboom

1. Introduction

The underlying fundamentals that have propelled the growth of global trade over the last decades beg to question their rationale and sustainability. The end of asset inflation, the decline in debt based consumption, the overreliance on export oriented strategies and the associated trade imbalances are imposing stringent readjustments on freight distribution systems and the global value chains they support. Yet, from a business cycle perspective, periods of growth are commonly followed by adjustment phases where misallocations are corrected, particularly if based on credit. Thus, the wave that has led to impressive growth figures in transport demand may shift towards a new paradigm that could have substantial consequences on the operating conditions of maritime shipping companies and transport terminals. This phase of readjustment is further exacerbated by the extended role that finance has taken in maritime shipping, trade and transport terminals in recent years.

The global economic crisis, triggered in 2008 by an unprecedented financial crisis, has taken on vast proportions. The crisis resulted in a generalized recession in all OECD countries and in many emerging economies, which is fundamentally challenging the direction of future trade flows and the sense of present trade organizational arrangements. Dependable factors (the stability of the world's financial institutions, continuous and sustainable GDP growth, the reliance on the backbone economies of the OECD) and unfailing certainties (government intervention in the economy, the superiority of widely applied logistics concepts) are put into question and contested, if not opposed.

Since the financial industry has taken such an active role in global economic affairs, understanding global trade and transportation requires, more than ever, insight into financial issues and their impacts on transport operations. Paradoxically, this insight is weak in the contemporary analysis of both maritime shipping networks and port economics. For instance, the strategies of maritime shipping companies and of port operators and the sensitivity of supply chains to cost variations are fairly well known processes that have helped explain how maritime transport systems adapt to and shape changes. Yet, this perspective sheds limited light on one of the fastest and most radical changes ever to affect the maritime and port industries. Since the economic crisis that began in 2008 initially concerned the financial sector, it is through the lens of financial issues that its consequences on the maritime industry can be understood. It is paradoxical that maritime transportation has become highly intertwined with the financial sector while its main drivers are not mainly financial, but macroeconomic issues.

To deal with the effects of changes in world trade, it is necessary to investigate more deeply the current and anticipated state of affairs regarding the world economy, value chains, the maritime transport industry, the ports, and the terminal operations sector in order to identify the root causes for their present situation. This chapter will, after an analysis of the evolution of world trade between 1990 and 2008, look at the reasons for the dysfunction of the world economy since late 2008 and in particular at the sectors mentioned above. Still, the process of globalization of trade that has surged since the 1990s gave numerous reasons to support the impressive growth predictions for the maritime and port industries.

2. Evolution of World Merchandise and Seaborne Trade

2.1 Unmitigated Growth in Global Trade

Looking at the evolution of international trade since the early 1990s underlines an unmitigated growth with short decline periods, such as during the Asian financial crisis of 1997 and the recession of 2000-2001. A convergence of factors supported this substantial growth. First, integration processes, namely various forms of regional and global trade agreements, promoted trade as regulatory regimes became better harmonized (e.g. tariffs). Examples include the accession of China to the World Trade Organization in 2001 and the creation of the European internal market in 1993. Second, production systems became more fragmented as it became easier to seek global comparative advantages in terms of labour and accessibility to markets, notably through globally scattered production sites and global sourcing strategies. Third, international transportation systems, maritime shipping and port terminals saw a substantial development in capacity, connectivity and reliability. Fourth, the transactional efficiency of international trade was improved with ever more sophisticated and high-performing telecommunications and information technologies as well as a greater availability of capital to finance international trade transactions.

It comes as no surprise that the growth of maritime transportation is strongly correlated with the growth of international trade, as maritime shipping and ports are the main physical support for international trade transactions. The value of global exports first exceeded USD 1 trillion in 1977 and by 2008, more than USD 16 trillion of merchandise was exported (Figure 2.1). During the same time period, the share of the world GDP accounted by merchandise

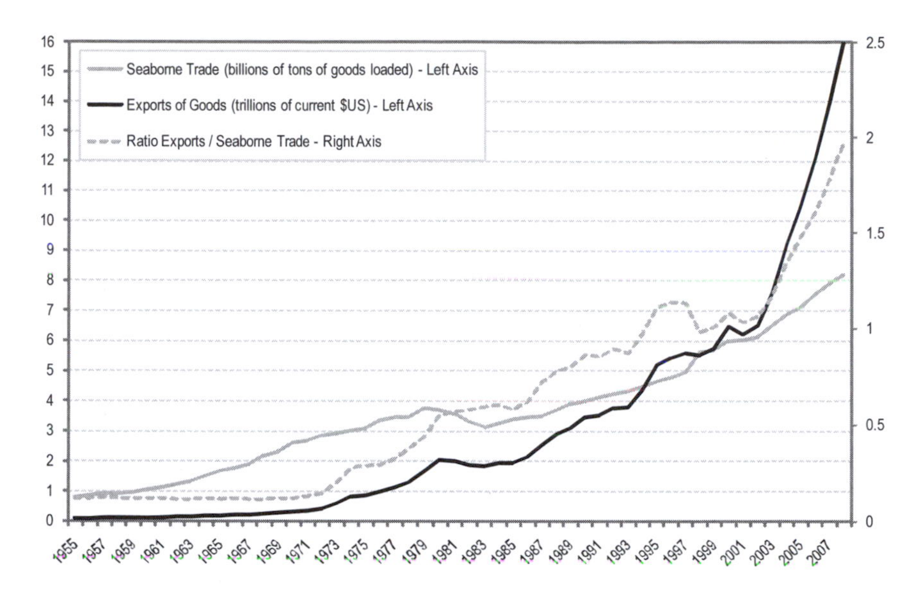

Figure 2.1 International Seaborne Trade and Exports of Goods, 1955-2008 (Data in current USD)

Source: WTO and United Nations, Review of Maritime Transport.

trade, imports and exports combined, surged from 18 per cent to 51 per cent. As expected, major fluctuations in the value of exports in the 1970s and 1980s were mainly linked with economic cycles. More recently, the development of containerized maritime transportation was linked to a growing trade of value-added commodities. From the late 1990s, a growing disconnect took place between the volume and value of maritime trade, mainly the outcome of the increasing sophistication of goods manufactured in Pacific Asia and rising energy (oil) prices as well as the spatial fragmentation of production, since components could be traded several times.

The ratio between value of exports and volume of seaborne trade remained constant until the first oil shock in 1973, underlining little change in the composition of maritime shipping. The first two significant changes in this ratio correspond to the first and second oil shocks, implying that higher oil prices directly impacted the global value of exports. Afterwards, the steady growth of the ratio is mainly attributed to the growth in the containerized trade of high value merchandise, particularly at the beginning of the twenty-first century.

2.2 Unparalleled Growth in Containerization ...

The world container throughput is the summation of all containers handled by ports, either as imports, exports or transshipment (Figure 2.2). This means that

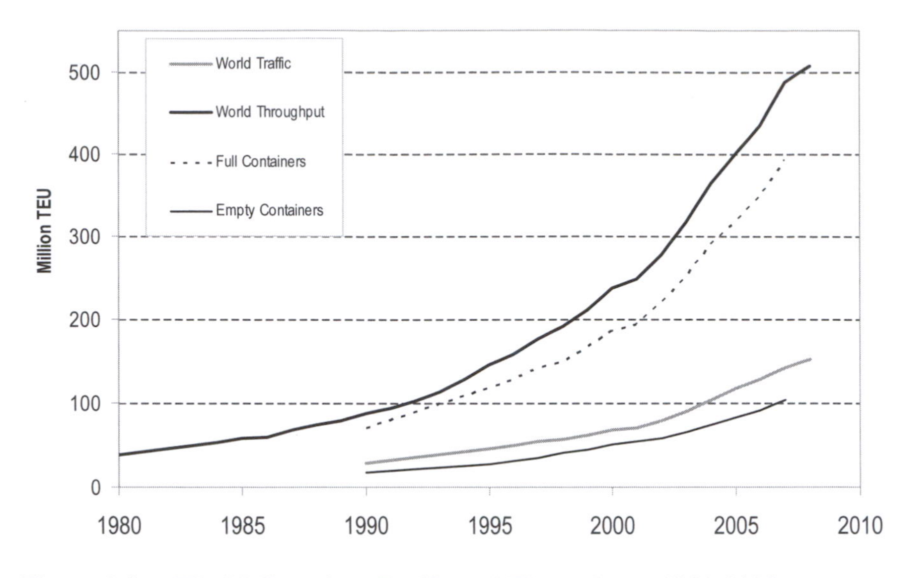

Figure 2.2 World Container Traffic and Throughput, 1980-2008
Source: Drewry Shipping Consultants.

a container is at least counted twice – as an import and as an export – but also each time it is handled at the ship-to-shore interface (e.g. at an intermediate location). Thus throughput should ideally be counted in container moves, but for basic commercial-strategic reasons, both port authorities and terminal operators prefer to communicate throughput figures in TEU. The world container traffic is the absolute number of containers being carried by sea, excluding the double counts of imports and exports as well as the number of transshipments involved. The throughput numbers reflect the level of transport activity while the traffic reflects the level of trade activity. The trend underlines a divergence between both as global supply chains and liner shipping networks (e.g. hub-and-spoke systems) became more complex.

Between 1990 and 2008, container traffic has grown from 28.7 million TEU to 152.0 million TEU, an increase by about 430 per cent. This corresponds to an average annual compound growth of 9.5 per cent. In the same period, container throughput went from 88 million to 530 million TEU, an increase of 500 per cent, equivalent to an average annual compound growth of 10.5 per cent. Consequently, the ratio of container traffic over container throughput was around 3.5 in 2008, whereas this ratio stood at 3.0 in 1990. The surge of both container traffic and throughput is linked with the growth of international trade in addition to the adoption of containerization as a privileged vector for maritime shipping and inland transportation. Temptation has been great for forecasters to extrapolate exponential growth for the next 15 to 20 years resulting in phenomenal throughput levels. Unfortunately, too many forecasts are exclusively developed from a top-down perspective; too few consider the physical cargo flows that external trade will be able to generate and the inevitable shifts in cargo types, routes and packaging. This also means that a small drop in traffic can result in a significant drop in throughput.

2.3 ... and the Surge of Trade Imbalances

An overview of the world's largest exporters and importers underlines a unique situation. Some countries, notably the United States and the United Kingdom, have significant merchandise trade deficits which are reflected in their balance of payments. Three main causes explain these trade imbalances:

- Comparative advantages and the search for lower input costs have provided powerful incentives to consider new locations for production. This enabled production costs to be kept low for a wide array of manufacturing activities, notably because of a "Chinese price" that became a frame of reference for labor intensive manufacturing activities. This greatly helped to offer consumption at cheap thus attractive prices, thereby fuelling demand in countries already facing serious trade deficits. Thus, the redistribution of production was a very unequal process, while comparatively consumption at a global level remained more stable.
- Foreign direct investments were the main means used to support comparative advantages and implied a transfer of capital and technology as well as its accumulation as infrastructure and means of production. This process combined with global sourcing strategies accelerated global trade, but also the setting of imbalances.
- Debt and asset inflation are a much less known cause, but are of prime importance. The growth of international trade, notably after 2002, was correlated with a phase of asset inflation, particularly real estate. The real estate bubble that took place in the United States and several other countries (most notably, the United Kingdom and Spain) between 2001 and 2006 was accompanied by a staggering growth of debt using assets as collateral. A share of this debt though was used to consume imported goods, impacting trade balances.

In 2008, five countries accounted for close to 70 per cent of global trade deficits: the United States, France, Italy, the United Kingdom and Australia. Conversely, countries having a positive trade balance tended to be export-oriented and dependent on international trade for their economic growth. Taken together, they generated 83 per cent of the global financial surplus in 2008 (IMF 2009). Economic history teaches that acute imbalances cannot be maintained indefinitely without a readjustment. This commonly emanates in a sharp and rapid correction.

3. The Economic Crisis Unraveled

3.1 Shifting Relations between Maritime Transportation, Trade and Finance

The links between the financial industry and maritime shipping are very old. Shippers have a long tradition of interaction with the financial industry as funding was required to build ships and to purchase trade cargo, while mitigating the risks related to shipping led to the creation of the insurance industry (for instance Lloyd's of London in 1871). The traditional role of the financial industry was a more passive one, providing capital and minimizing risk when needed. This capital was often paid back once a voyage was completed and the cargo sold. Finance was used to leverage the opportunities of international transportation. Yet in the last decade this relationship has become more acute because it was inverted. Indeed, transportation became a means to leverage financial opportunities for the following reasons:

- Transport nodes and terminals are very capital intensive. The substantial levels of productivity brought by containerization resulted in a much more capital intensive industry depending on financing for the acquisition of maritime assets. The amortization of investments tends to take place over longer periods of time. In the past, the vast majority of port infrastructures were owned by public bodies. In the 1980s, and accelerating through to the new century, governments around the world sought to divest themselves of the financial and operational burdens of ports and began to lease terminals to international terminal operators. These new operators began to attract financial support from banks, insurance companies and pension firms. This provided large quantities of capital to develop intermodal assets and an increase of their "spot" value.
- Financing international transactions. With the growth of international trade, transactions between commercial actors became increasingly complex and reliant on financing. Letters of credit are mainly used for transactions between actors, such as a buyer and a seller, in different countries. Large commercial banks commonly finance about 90 per cent of all the global trade transactions (Leach 2008).
- Shipping derivatives. During the last decade, the maritime and port industry became increasingly intertwined with the financial world. The high volatility in the shipping markets, exemplified by sharp fluctuations and sudden changes, has supported the emergence and growth of a paper market for shipping freight. Complex financial products and derivatives have been developed to support the growth in shipping (Kavussanos and Visvikis 2006). Shipping derivatives have been developed in order to manage risks emanating from fluctuations in freight rates, bunker prices, vessel prices, scrap prices, interest rates, and foreign exchange rates more effectively, in a cheaper and more flexible manner. The shipping market now makes extensive use of risk management techniques and instruments

> attracting trading houses and energy companies, as well as investment banks and hedge funds. The risks, if managed effectively, can stabilize cash flows, with positive repercussions for business.

Until late 2008, these issues substantially benefited the maritime and ports industry as large sums of capital became available for a variety of intermodal improvements such as a new generation of containerships, the development of new and more advanced port terminal facilities worldwide to support export-oriented strategies (as illustrated by those in East Asia), and inland ports to better access regional markets (for example in North America and Europe). The increasing involvement of the financial industry however led many actors in the shipping industry to expand their fleets (see "Maritime Shipping and Ports in a Flux", this chapter). Furthermore, transportation became increasingly perceived solely from a financial perspective, particularly since a large number of stakeholders and decision makers were coming from the financial sector as opposed to the transportation sector. Terminals became a financial product part of a global asset portfolio, whose performance was often seen in terms of price to earnings ratios. Another perverse effect was that terminal assets came to be perceived as liquid, a perception encouraged by the active involvement of a variety of financial firms so that terminal assets for sale (or lease) could readily find an acquirer. The problem is that financial considerations can shift rapidly as the horizon is commonly short term, while intermodal assets have a planning and operational horizon that can easily span a decade even for its most "volatile" elements such as ships and several decades for its less "volatile" elements such as terminal infrastructure. The volatility that characterizes financial markets permeated the maritime shipping industry, and as the global economy surged under a flood of cheap credit coupled with asset inflation, so did the shipping and port industries.

3.2 The Underlying Dynamics behind the Economic Crisis

The world reserve currency status of the Dollar conferred to the United States much flexibility in its monetary policy with the capacity to issue large amounts of debt without raising much concern from its creditors. While central banks have a level of control over interest rates, they have limited if any control over the economic sectors that accumulate the credit they create. After the stock market contraction of 2001 the Federal Reserve lowered interest rates. Credit accumulated in the real estate sector, thereby creating a very large speculative bubble. Facing rapid inflation of their real estate assets, millions of consumers contracted additional debt, which otherwise would not have been possible. A significant share of this debt went into consumption.

Many trade partners, mainly in Asia, took part in the accumulation of this debt by financing it. A share of the dollars traded for goods has accumulated as monetary reserves and served as investment capital for production and distribution infrastructures across the world. China and other Asian export-oriented nations understand well that no markets other than the United States and the European Union are large enough to consume the flows of goods they generate. It was in their strategic interest to prop foreign consumption and find venues to invest trade surpluses, particularly in a period when China's domestic demand was not high enough to sustain a strong economic growth pattern. A share of the positive balance of payments thus came back to the United States in the form of the purchase of financial assets (treasuries, mortgage backed securities, bonds, and so on), supporting the dollar and indirectly favoring the continuation of the real estate bubble and the consumption-derived debt it created. This process could, however, not endure forever and thus the stage was set for one of the largest financial collapses in history.

3.3 Enter the Crisis

The economic crisis of 2008-2009, induced by the huge toxic debts of financial institutions, had fundamental consequences on international trade. Many explanations can be offered as to what caused this rapid and unprecedented collapse of the main world economies. Following are some of the more obvious and less contentious:

- The huge balance of payments deficit of the US and the resulting weakness of the US dollar that was mitigated by large purchases of American debt instruments by foreign financial institutions;
- The deflation of real estate assets (e.g. massive defaults in sub-prime loans) resulting in a full-fledged financial disaster for major banks around the world as they were forced to re-price underlying assets to levels triggering the insolvency of many;
- Overstocking induced by cheap credit and generous rebates and the expectation in 2007 that inflation would sharply rise following massive increases in oil prices;
- Creation of excess capacities in many countries and sectors, following the artificially sustained boom, helped by low interest rates available to borrowers.

Each of these causes can on its own bring about a serious economic decline, but it is their inter-relationship and combined impact that have pushed all major economic indicators (GDP growth, trade growth, employment) into negative values. The above observations thus underline that global trade was built in the late 1990's and up to 2008 under unsustainable economic and financial foundations.

3.4 Reaping the Macroeconomic Storm that was Sown

The accumulation and the subsequent defaults on massive amounts of debt played a fundamental role in the financial crisis. Bubbles, which are nothing more that credit-driven booms, give wrong signals to the economy by misguiding investment and capital accumulation processes since fundamentals are distorted and appear more significant than they really are. More capital than required is accumulated in activities related to the bubble, creating a hidden overcapacity linked to an artificially induced peak in demand. In North America and a number of European countries the credit bubble resulted in an overcapacity in residential and commercial real estate and created an artificial level of consumption. For export-oriented economies, overcapacity took place in the setting of production and distribution assets incited by the debt-derived growth of consumption taking place because of asset inflation.

Towards the end of 2008, credit suddenly became scarce as many debt instruments, particularly those related to real estate, started to experience excessive default levels, beyond anything expected in risk assessment models. This created massive cross-defaults within an over-leveraged financial system and forced the downward re-pricing of whole asset classes, pushing many institutions into insolvency, unable to service the massive debt obligations they contracted under the assumption that the underlying assets were liquid and rising in value. Many financial institutions became excessively unwilling to lend, uncertain about whom would be next to default while experiencing a deflation of their own assets and balance sheets. The outcome was a self-reinforcing vicious circle that began in the financial sector and rapidly spilled over in the material economy, triggering a global recession that began to be felt in the fall of 2007 and a global collapse of equity markets a year later. Recessions have substantial impacts on the freight transport sector, as articulated in Figure 2.3.

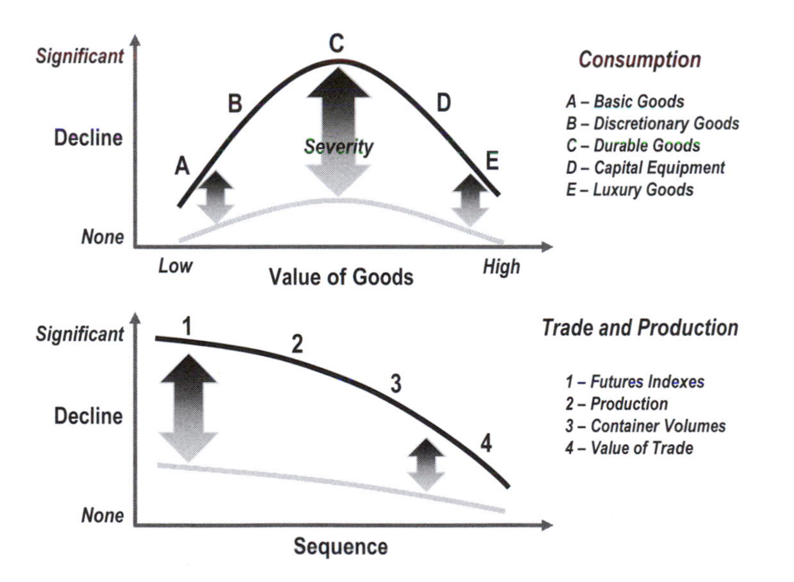

Figure 2.3 Potential Impacts of Recessions on Trade and Freight Transportation by Level of Severity (Low and High Scenarios)

The impact level of a recession on many sectors of the economy involves downward pressures on prices in view of collapsing demand. Basic goods (e.g. food) tend to be the most resilient, so their respective supply chains tend to be impacted to a lesser degree by recessions. However, it is over durable goods (e.g. cars), discretionary goods (e.g. electronics) and capital equipment (e.g. ships, port infrastructure), that recessionary forces can have significant impacts in lowering their respective levels of consumption. It is precisely over these product classes that export oriented economies have focused their development strategies. Stock market valuations and freight rates (futures indexes) tend to be leading forces in the decline of international trade, followed by income, spending and seaborne freight (particularly container volumes). What is notable for the correction that began in 2008 is the extreme rapidity at which the sequence unfolded, implying that while future (forward looking) indexes first collapsed, so did container volumes and global trade immediately afterwards, confirming the inevitability of the collapse of the material economy and the temporary and for some freight segments the permanent disappearance of substantial portions of merchandise volume.

4. Maritime Shipping and Ports in a Flux

4.1 Future Indexes, Shipping and Trade: Falling Off a Cliff

The Baltic Dry Index (BDI) is an assessment of the average price to ship raw materials (such as coal, iron ore, cement and grains) on a number of shipping routes and by ship size (The Baltic Exchange 2009). It is thus an indicator of the cost paid to ship raw materials on global markets and an important component of input costs. As such, the index is considered as a leading indicator (forward looking) of economic activity since it involves events taking place at the earlier stages of global commodity chains. A high BDI is an indication of a tight shipping supply and is likely to create inflationary pressures along supply chains. A sudden and sharp decline of the BDI is likely to foretell a recession since producers have substantially curtailed their demand leaving carriers to substantially reduce their rates as maritime capacity cannot by rapidly reduced.

Between mid-2005 and mid-2008 the BDI grew by a factor of about 5.5 times, reflecting an almost surreal surge in global trade and expectations of additional growth, mainly fuelled by a Chinese economy hungry for raw materials and energy. The shipping industry was increasingly facing limited extra capacity and port congestion or the expectation of congestion (UNCTAD 2009; Notteboom 2006a). The existing capacity shortages in vessels and terminals pushed rates up to unparalleled heights. The index peaked in the spring of 2008 as China was stockpiling large quantities of commodities in preparation for the Olympics. Afterwards, the BDI reflected the full force of the unfolding recession and collapsed by 94 per cent between July and December 2008 (Figure 2.4). Never

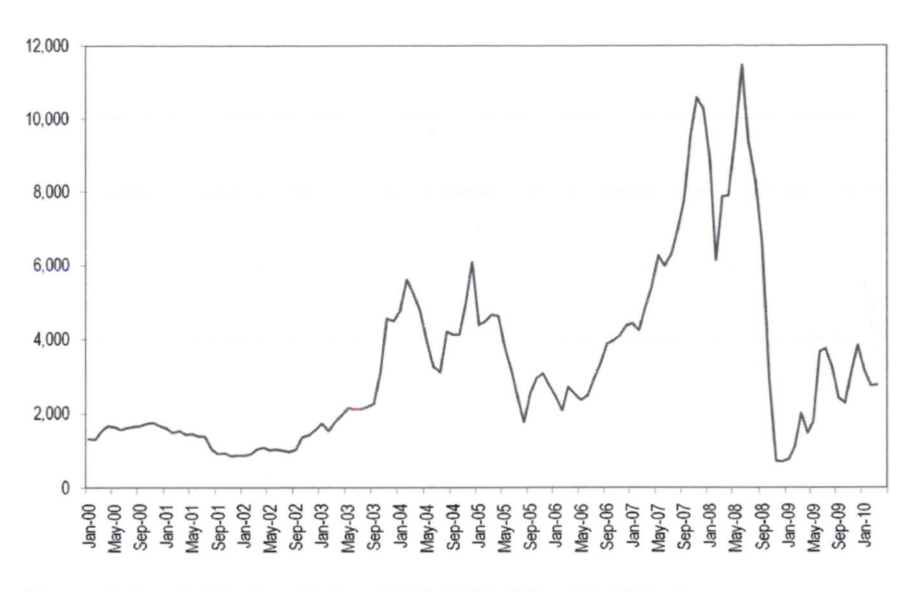

Figure 2.4 Baltic Dry Index, 2000-2009 (Monthly Value)
Source: The Baltic Exchange.

before was such a sharp correction observed, an indication that maritime shipping and global trade was brought to a full recession. Subsequently, the BDI corrected to attain a level reflecting pre-bubble valuations.

The collapse of global shipping observed in the BDI, which was further confirmed by the simultaneous downfall of container freight rates on all major routes, is the outcome of two consecutive storms:

- The credit storm: In the fall of 2008 the credit freeze implied that several financial transactions in international trade could no longer be cleared under normal conditions. For instance, the terms to obtain a letter of credit became much more stringent, with some financial institutions refusing to honor letters of credit issued by foreign banks and others simply refusing to issue credit to their customers perceived to be at risk (Leach 2008). Therefore, containerized trade volumes started dropping precipitously because transactions could not be cleared. Access to credit was thus a temporary explanation in the collapse of maritime shipping as several transactions between developed and developing countries were perceived as risky and could not be financed, at least in terms traders were accustomed to. The weakness of the financial sector, particularly in developing countries that have become important traders, compounded the collapse. However, this was a temporary condition reflecting a sudden change in the global financial industry as it absorbed a new reality, assessed risk and provided financing accordingly.

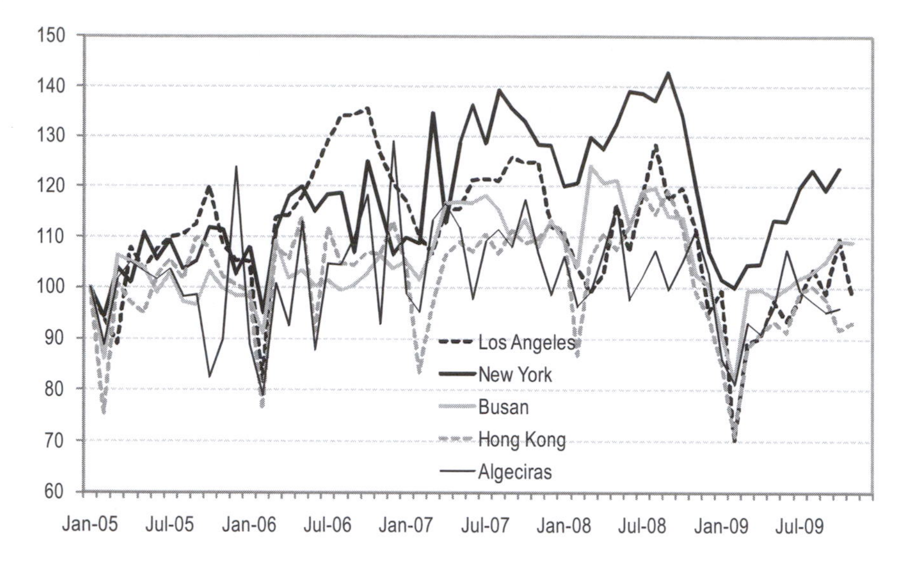

Figure 2.5 Monthly Total Container Traffic at Selected Ports (Jan 2005=100)
Source: Websites of Respective Port Authorities.

- The macroeconomic storm: The downward pressure on maritime shipping is linked with a decline in aggregate demand, which is a standard macroeconomic impact of a recession. Consumption in North America and Europe is unlikely to recover until consumption is more synchronized with income, as opposed to debt accumulation. There are many forecasts made about the timing of the expected economic recovery, but these forecasts are constantly being revised downward and recovery is pushed further up the calendar. The credibility of analysts that did not foresee or even anticipate the possibility of the largest financial and economic correction since the Great Depression begs to place limited value on their expectations about the potential timing and scale of the recovery. The compounding medium term effects of an emerging pension crisis, the aging of the global population in the traditional industrialized countries and serious levels of debt defaults are other factors to be added.

Both of these storms have readily been felt in international trade and container port volumes (Figure 2.5) as in the fall of 2008 there was a systematic and correlated drop among all major trade partners and gateways.

The pattern observed in Figure 2.5 underlines a reassessment of neo-mercantilism trade policies, particularly for export-oriented economies. While the financial storm had a sharp and sudden impact on volumes, the macro-economic storm will span several years and will involve several phases of growth and decline. It will endure until trade and macroeconomic imbalances have been substantially rectified.

Table 2.1 Ships on Order and in Service on 1 July 2008 by Main Class of Ship

Ship type	Tankers	Bulk carriers	Container ships	General cargo ships
No. of ships on order	2,956	3,118	1,451	1,714
Total dwt (1,000 tons)	193,652	271,713	82,185	22,095
Total no. of ships in world fleet	11,525	7,357	4,475	17,756
Total dwt in world fleet (1,000 tons)	404,891	401,949	154,396	105,101
Ratio ships on order to ships in world fleet	25.6%	42.4%	32.4%	9.6%
Ratio dwt on order to total dwt of world fleet	47.8%	67.6%	53.2%	21.0%

Source: ISL Bremen.

4.2 A Double Squeeze between Supply and Demand

Maritime shipping benefited from the prevailing growth in traffic and in particular from the almost unlimited supply of credit for building new vessels. Table 2.1 shows the order book for major classes of ships as of 1 July 2008 just before the world economic crisis started to unfold. The ships and deadweight on order are then compared with the total number of ships and the total deadweight in service at that date. This comparison confirms the magnitude of the order book in particular for bulk carriers and containerships, with orders for new such ship types totaling respectively 67.6 per cent and 53.2 per cent of existing capacity.

For cellular container ships the level of extra capacity that was planned to be added can equally be illustrated by the number of ships and TEU capacity for each vessel size. The cellular container ship order book as of 1 August 2008 is shown on the left side of Figure 2.6. There was an expected increase of 36 per cent in number of ships and of 67 per cent in carrying capacity expressed in number of TEUs (equivalent to a compound growth rate of 14.7 per cent per annum). Moreover, the greatest number of additions to the fleet was expected to occur in the ship classes with capacities of over 5,000 TEU.

The economic crisis faces the shipping sector with a double threat. The first is posed by a significant drop in demand as a result of the serious slump in global trade witnessed worldwide. Not all goods and all ship types feel this slump in the same proportions. Very seriously affected are all classes of bulk carriers and most classes of cellular containerships. Tankers and in particular VLCCs have until recently been much in demand, but other tanker types and classes have had a rough ride in early 2009.

The second threat comes from ship owners massively ordering new ships for which they believed, at the time the orders were placed, that there would be much demand. The ease with which ship financing was possible up to the second half

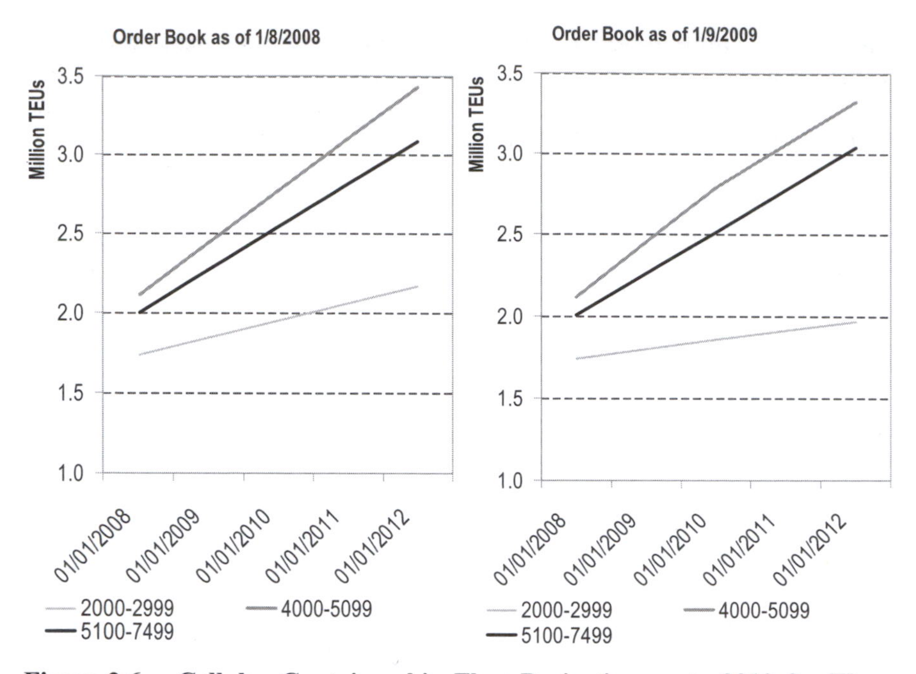

Figure 2.6 Cellular Containership Fleet Projection up to 2012 for Three Selected Classes

Source: AXS Alphaliner (2009).

of 2008 and the low interest rates induced many to go into investments for which demand had been unproven. But the prevailing mood at the time was one of over optimism based on the belief that world trade would continue to grow without relenting. This optimism prevailed in the forecasting of future shipping volumes and their underlying growth rates.

By September 2009 container line operators had reassessed the market situation and introduced a host of measures to reduce capacity, including delaying delivery dates of ordered vessels, returning chartered vessels to their owners and canceling when possible new-building orders. The right side of Figure 2.6 shows the cellular containership fleet projection but based on the fleet in service on 1 September 2009 and the revised order book at the same date. It underlines that the greatest changes in total capacity are observed for ships below 5,000 TEU.

A comparison of the expected fleet figures on 1 January 2010 (based on the fleet on 1 September 2009) with the fleet figures on 1 August 2008 shows the extent of capacity control exerted by the container ship owners and container line operators. Total number of ships by 1 January 2012 is expected to be down by 437 whilst TEU capacity will be 1.476 million TEU less. Although these are significant reductions, the fleet is still to grow by 24 per cent in TEU capacity in the two-year period 1 January 2010 to 1 January 2012. Without a further strong reduction in number of vessels and TEU capacity of the fleet, there can be little doubt that serious overcapacity will remain a major problem for the container shipping industry, even if containerized trade volumes would come back with moderate growth over and above 2008 levels.

4.3 Revisiting Forecasting and Port Growth Assumptions

The problem of short-, medium- and long-term forecasting of the demand for shipping is not new, but made particularly acute in the current context. In the past, decisions were made by people directly involved in shipping, many with sea-going experience and solid connections in the maritime community. Today, some of the decision makers in shipping have a non-maritime background, mostly from finance. They frequently have different viewpoints and objectives from those involved in the day-to-day running of shipping services. They tend to defend the interests of specialized maritime lending institutions, merchant bankers or shareholders which have limited understanding of the shipping business and mainly look for short-term returns on their invested capital instead of trying to anticipate long-term cycles in shipping markets.

There remains the uncertainty of future demand for shipping. Many forecasts are prepared (frequently by bankers and economic advisers) without giving much attention to the physical cargo flows that are the generators of the demand for shipping services. This is particularly evident in the forecasting of container volumes, whereby these forecasts suffer from at least two major flaws:

- The disconnection between the physical cargo flow (what cargo is carried in the box) and the top-down forecasting of number of boxes inevitably induces parties with a vested interest to bend the container figures in their favor. This is the more frequently done if no regard has to be paid to real physical cargoes to be carried;
- The confusion that continues to exist with respect to the notion of seaborne traffic (what is carried by sea in international trade) and that of throughput (the effort required in ports or terminals to handle the seaborne traffic). Many studies, reports and miscellaneous publications present maritime or seaborne traffic values whilst effectively referring to port throughput values.

Thus, forecasting future container port volumes, while in the best of times a risky proposition, has been compounded in recent years by several factors that may have further contributed to extrapolation fallacies and led to overestimations of future traffic. By using a sample of large ports, Table 2.2 underlines that using compound annual growth (CAG), a commonly used method in finance and business, for forecasting container volumes can lead to staggering overestimations of future growth potential. Such an approach is commonly justified by the growth pattern that took place since the early 1990s and which effectively mimic this behavior until 2007. Yet, using CAG for port volumes forecasting is a dangerous fallacy.

Table 2.2 2020 Throughput Forecast, Selected Ports, Linear and CAG Scenario

Port/Traffic 2007	R2/CAG (1998-2007)	Traffic 2020 (Linear Scenario)/CAG	Traffic 2020 (CAG 1998-2007 Scenario)
New York/5.3	0.996/+7.9%	9.6/+4.7%	14.2
Savannah/2.6	0.968/+13.5%	4.9/+5.1%	13.6
Los Angeles/8.3	0.966/+9.5%	16.6/+5.4%	27.1
Antwerp/8.2	0.974/+9.6%	14.5/+4.5%	26.9
Algeciras/3.4	0.961/+6.5%	6.0/+4.4%	7.7
Busan/13.3	0.983/+8.4%	24.3/+4.8%	38.1
Shanghai/26.1	0.948/+23.9%	56.5/+6.1%	423.8
Montreal/1.4	0.944/+3.8	1.9/+2.8%	2.2

Note: All traffic figures are in million TEU. CAG: Compound Annual Growth. Forecast does not consider port capacity constraints.

The following observations can be made:

- A linear regression analysis of a 10-year period between 1998 and 2007 with volume as a dependent variable and year as an independent variable reveals that all ports have a very high coefficient of determination (R^2) of their annual throughput. This implies that the growth pattern for container throughput was mostly linear for that decade and would thus incite using linear forecasting methods to assess future volumes. For instance, for the port of New York, throughput growth between 1998 and 2007 took place in a nearly perfect linear fashion (R^2 of 0.996), which corresponds to a CAG of 7.9 per cent for the same time period.
- Throughput differences between the linear scenario and the CAG scenario are substantial. While the linear scenario reflects well past growth figures, as evidenced by the very high coefficients of determination, CAG scenarios abound in the port traffic forecasting literature. The case of New York underlines that for 2020 the linear growth expectations would place the throughput at 9.6 million TEU, while using the observed CAG of the period 1998-2007, standing at 7.9 per cent, would place throughput at 14.2 million TEU, a difference of 4.6 million TEU. This simple difference in forecasting methodology for a period of 13 years (2007 to 2020) accounts for close to the existing volume of the port. The case of Shanghai is absolute and obviously completely unrealistic, as the CAG scenario based on the fast growth figures between 1998 and 2007 (23.9 per cent), would place the port with a throughput in 2020 similar to the global container port throughput in 2005.
- Forecasts are commonly done by port authorities (or hired consultants) and tend to have a "pro growth" bias. This bias assumes that growth implicitly comes from new volume and that port competition and alternate shipping

options are not at play. Sometimes, the fast growth taking place in a sample of large ports is taken over by smaller ports as part of their business plans. For instance, the Montreal 2020 plan was expecting a throughput of 3.6 million TEUs by 2020 (from 1.4 million TEU in 2006), which accounts for a CAG of 7 per cent, double the 3.8 per cent CAG rate observed between 1998 and 2007 for that port.

There are many shortcomings behind using CAG for container volume forecasting. First, CAG assumes a steady growth rate where volatility and market changes cannot be effectively captured. Second, CAG is based on a past time series that does not necessarily warrant future expectations. Last, and not least, the time series used to calculate CAG can be selected to skew results according to preferences. The time series of Table 2.2 (1998 to 2007) were deliberately selected to correspond to the most significant growth in container port volumes. Therefore, the issue is not the errors compounded by the forecasting methodologies, as it remains just a quantitative exercise, but the misallocation and overinvestment that they incite can lead several segments of the maritime industry in dire financial strain. It is evident that during the decade prior to 2007-2008, growth in the maritime shipping and port industries has incited the most positive forecasting prospects and that methodologies can be very easily bent to rationally provide astounding figures. This perspective can no longer prevail and a new reality for ports and maritime shipping will need to be the order of the day.

5. Conclusion: Rediscovering Reality

The unprecedented economic downturn is fundamentally challenging the direction of future trade relations and their corresponding physical flows. From this paradigm shift in globalization lessons can thus be drawn with respect to traffic and throughput forecasting. First it would suit all those that try to develop prognoses of future activity to be extremely modest. The likelihood that the pre-recession estimates really will be achieved is virtually nonexistent and there are too many factors influencing the results. After all, seaborne traffic and port handling are derived demands that are a consequence of their insertion in trade systems, a factor that has frequently been overlooked in a recent past. It is as if any additional capacity would be fulfilled with a corresponding demand. But one has also to acknowledge that even a general economic downturn has very different effects in various parts of the world.

The second lesson from recent events is the risk of relying both on perception (based on statements such as "there is a general belief that", "the consensus between stakeholders is" etc.) and on general but unproven statements ("de-stocking is now over and traffic will pick up again" etc.) rather than on hard facts and a close monitoring of developments at the grass root level. Moreover, there are far too many forecasts based on complex but unproven models and far too few that start from painstaking and difficult bottom-up research on individual value chains.

The third lesson is one of realism. Those that are closest to the trade have the greatest interest to keep their information confidential and possibly to circulate a certain level of disinformation. This Achilles' heel is further exposed by a complete lack of uniformity and conformity in reporting by various international organizations and sector specialists. It makes data comparison extremely precarious if not misleading.

The impact of the global economic downturn, the crisis in the main shipping sectors and the changes in the pattern of world trade all profoundly affect the maritime transport industry. The resulting consequences for shipping in general, and ports and trade corridors in particular, have both long-term and short-term policy and strategy implications. These will be considered in more detail in the next chapter.

Chapter 3

Organizational and Geographical Ramifications of the 2008-2009 Financial Crisis on the Maritime Shipping and Port Industries

Theo Notteboom, Jean-Paul Rodrigue and Gustaaf De Monie

1. Introduction

The previous chapter questioned the underlying fundamentals that have propelled the growth of global trade over the last decades. The present downturn, after years of fast economic growth, is fundamentally challenging the direction of future trade flows and the sense of present trade organizational arrangements. From a business cycle perspective periods of growth are commonly followed by adjustment phases where misallocations are corrected, particularly if these are based on credit. This chapter explores the ramifications that the readjustment of global imbalances entails over three issues:

- The first ramification concerns the impact of the crisis on strategies of container carriers and terminal operators, in particular their operational strategies with respect to vertical and horizontal integration.
- The second ramification involves how the forces behind the added value creation process are reconciled with the debasement of value chains through competition and efficiency improvements. While the value capture process is a much sought after effect, particularly by gateways between major systems of circulation, to what extent can this value be diluted by diminishing returns?
- The third ramification pertains to the duality between concentration versus deconcentration, particularly in terms of ports, gateways and corridors. In an environment where traffic is likely to consolidate would additional concentration at major gateways and corridors accelerate as shipping lines rationalize their services?

The above issues combined should provide an answer to the pressing question: is a turning point in the making in global freight distribution and value chains, or are we witnessing nothing more than a temporary market correction that will not have a major impact on the logistics trends of the past few years before the crisis of 2008-2009 occurred?

2. First Ramification: The Strategies of Shipping Lines and Terminal Operators

During the past decennia, shipping lines and terminal operators have tailored their business strategies under the premises of strong growth in container trade, fuelled by globalization and the large-scale adoption of the container. Shipping lines and terminal operators have benefited greatly from these developments. However, the economic crisis seems to have shaken the fundamentals of the pricing and investment strategy of shipping lines and terminal operators and their broader involvement in value chains.

2.1 The Dynamics Driving the Liner Shipping Industry

Notteboom (2004a) provides an extensive overview of the dynamics behind the liner shipping business. Shipping remains a very capital-intensive industry where some assets are owned and others are leased and there exists a wide variability in cost bases (Brooks 2000). Once the large and expensive liner networks are set up, the pressure is on to fill the ships with freight as the economies of scale they provide can be brought to bear in terms of pricing, capacity and frequency. Lines accept that they have to take whatever price is offered in the market. This acceptance has, in turn, led to intense concentration on costs.

Consecutive rounds of scale enlargements in vessel size (see Drewry Shipping Consultants 2001; Cullinane *et al.* 1999) have reduced the slot costs in container trades, but carriers have not reaped the full benefits of economies of scale at sea (Lim 1998). Adding post-panamax capacity gave a short-term competitive edge to the early mover, putting pressure on the followers in the market to upgrade their container fleet and to avert a serious unit cost disadvantage. A boomerang effect eventually also hurt the carrier who started the vessel upscaling round.

Shipping lines also rely on organizational scale increases. Horizontal integration in liner shipping comes in three forms: trade agreements such as liner conferences (which are outlawed in Europe since October 2008), operating agreements (e.g. vessel sharing agreements, slot chartering agreements, consortia and strategic alliances) and mergers and acquisitions. The economic rationality for mergers and acquisitions is rooted in the objective to size, growth, economies of scale, market share and market power. Cooperation between carriers serves as a means to secure economies of scale, to achieve critical mass in the scale of operation and to spread the high level of risk associated with investments in ships (Ryoo and Thanopoulou 1999; Slack *et al.* 2002). Alliances provide its members easy access to more loops or services with relative low cost implications and allow them to share terminals and to cooperate in many areas at sea and ashore, thereby achieving costs savings in the end.

In a shipping industry already dominated by large vessels, mergers, acquisitions and strategic alliances the potential cost savings at sea (marginal improvement) still left are getting smaller. Inland logistics is one of the most vital areas still left to cut costs. Besides cost and revenue considerations, the demand pull force of the market is a main driving force for carriers to integrate their services along the supply chain (Slack *et al.* 1996; Heaver 2002). Lines that are successful in achieving cost gains from better management of inland and container logistics can secure an important cost savings advantage.

2.2 The Reaction of the Liner Shipping Industry to the Downturn

Container lines have to a certain extent adjusted their strategy to cope with the recent significant drop in volumes. The main fields of action relate to capacity strategy and pricing strategy. The pursued capacity and pricing strategies have an impact on the overall market structure.

2.2.1 Capacity deployment The current crisis and overcapacity in the shipping industry is partly the result of exogenous factors such as a decrease in demand and partly the result of endogenous factors such as wrong investment decisions by shipping lines. Also, Randers and Göluke (2007) argue that the turbulence in shipping markets is partly the consequence of the collective action of the members of the shipping community. Much of the overcapacity problems are linked to the complete failure by the key stakeholders, in the first place the ship owners and providers of ship finance (many without a maritime background), to correctly anticipate the future markets for different ship types and sizes. Shipping lines have had to adjust their capacity deployment strategies. Until early 2008, shipyards still struggled to satisfy demand for new and bigger ships. However, this development has stalled. Due to the economic slowdown, a large surplus of cargo capacities emerged particularly on the Europe-Asia and Transpacific routes. As shipyards were still completing the numerous orders from previous years, total slot capacities in the market would continue climbing until 2012. In late 2008, a number of shipping lines started to postpone orders and older ships were put out of service in large numbers. In mid-April 2009, the worldwide laid-up fleet totaled about 1.3 million TEU or 10.4 per cent of the world container fleet. This figure decreased slightly to 9.9 per cent of the world fleet capacity in February 2010, with 508 laid-up ships totaling 1.3 million TEU (Alphaliner 2010). Most of the idled ships are of the class between 1,000 and 3,000 TEU capacity, while hardly any post-panamax vessels were laid-up.

Shipping lines massively suspended liner services particularly on the Far East-Europe and transpacific trade routes. Total capacity on the Far East-Europe trade fell by 21 per cent between October 2008 and March 2009 (Table 3.1).

Vessel lay-ups, order cancellations and service suspensions were not the only tools used by shipping lines to reduce capacity (Table 3.2). Many vessels continue to slow steam at around 18 to 19 knots, despite the cheaper bunker prices, as the

Table 3.1 Far East – Europe Capacity Situation

	March 2009	October 2008	% change
Total no. of weekly services (North Europe/Med)	45 (26/19)	64 (36/28)	-30%
Total ships deployed	406	549	-26%
Average vessel size (TEU)	7,310	6,517	12%
Total capacity (TEU)	2.97 million	3.58 million	-17%
Average weekly capacity (TEU): March 2009 vs. October 2008	319,301	405,901	-21%
Average weekly capacity (TEU): 1Q 2009 vs. 4Q 2008	335,793	397,350	-15%

Source: Notteboom (2009a) based on Alphaliner data.

Table 3.2 Changes in Fleet Operations, TEU-mile Supply (Estimates Made in October 2009)

	2006	2007	2008	2009 (est.)	2010 (est.)	2011 (est.)
Deliveries of new vessels	+14.1%	+13.1%	+12.7%	+15.6%	+23.6%	+20.4%
Delayed deliveries from previous years	-	-	-	+1.2%	+5.0%	+5.0%
Scrapping	-0.3%	-0.2%	-0.9%	-2%	-2%	-1%
Newbuilding delivery deferrals	+1.0%	-0.3%	-1.2%	-5%	-5%	-5%
Newbuilding cancellations	-	-	-	-2%	-5%	-10%
Lay-ups/service suspensions	-	-	-	-10%	-5%	-2.5%
Slow-steaming/re-routing	-	-1.9%	-5.6%	-7.5%	-5%	-5%
Effective Supply Growth	+14.8%	+10.7%	+5.0%	-9.7%	+6.6%	+1.9%
Effective Demand Growth	+12.4%	+11.3%	-0.3%	-12.4%	+10.1%	+4.6%

Source: Authors' compilation based on figures from Drewry Shipping Consultants (2008) and Goldman Sachs (2009).

longer roundtrip time helps to absorb surplus capacity in the market (i.e. more vessels needed per loop). In early 2009, Maersk Line and the Grand Alliance were examples of shipping lines temporarily opting for the Cape route around South Africa instead of the Suez Canal route, mainly on the Eastbound leg of the roundtrip. The Cape route has longer transit times, but shipping lines could avoid high toll fees on the Suez canal (up to nearly USD 700,000 for the largest vessels – one way) which are difficult to justify given the economic climate, cheap bunker fuel in the beginning of 2009 and poor ocean freights on return Asia routes.

The situation in the charter market is particularly interesting. In mid-2009 the world container slot capacity amounted to 12.4 million TEU. Containership operators own 6.2 million TEU of the global fleet, while the remaining half is owned by financiers through leasing contracts. Given the current market situation chartered-in vessels are returned when leases expire and, consequently, taken out of the market. A certain portion of the charter expiries were renewed with daily rates in October 2009 about 75 per cent lower than the average level of 2008. Many operators return chartered ships when leases expire, because they have their own newly built deliveries coming to market. This has a substantial impact on the balance sheet of those who have provided financing to these charter ships. Therefore, the market could see fewer container ships on trade routes, at least until rates rebound to profitable levels.

2.2.2 Pricing strategies A second shipping line strategy to the decline in shipping volumes is an adjustment of pricing strategies. In an environment of overcapacity, high fixed costs and product perishability, lines will chase short-run contributions filling containers at a marginal cost only approach, often leading to direct operational losses on the trades considered. Even though lower rates may allow carriers to take on extra cargo, it also reduces their profitability. In many cases, shipping lines can earn more money with higher rates and lower utilization than with lower rates and higher utilization.

One of the main influencing factors affecting the recent pricing strategies of shipping lines was the dramatic evolution in the bunker cost, one of the main operating costs for shipping lines (Notteboom and Vernimmen 2009). The first half of 2008 saw a steep increase in bunker costs, with heavy fuel oil peaking at around USD 700 per ton in July 2008, compared to about USD 200 in January 2007. The economic slowdown initially put a strong downward pressure on the fuel price, with bunkers in Rotterdam plummeting from USD 700 per ton in July 2008 to USD 171 by late December 2008. While this had a positive impact on vessel operating costs, shipping lines soon realized that filling their ships became extremely difficult with freight rates no longer generating enough revenue to cover operating costs. Spot container freight rates were reduced to virtually zero, with only a small compensation for fuel surcharges. Rates bottomed out in February/March 2009 as they could not go much lower (see the example in Table 3.3). In the second quarter of 2009, vessel capacity reductions on the major trade lanes started

Table 3.3 Base Freight Rate and Bunker Adjustment Factor (BAF) for the Maritime Transport of one 40 Foot Container (FEU) from Shanghai to Antwerp (Excluding CAF, THC and Other Surcharges)

	Typical freight rate (in USD)	Typical BAF (in USD)
Q1 2007	2,100	235
Q2 2008	1,400	1,242
September 2008	700	1,440
February 2009	750 (all in)	-
April 2009	1,500 (all in)	-
September 2009	2,200 (all in)	

Source: ITMMA based on market figures.

to have a positive effect on rates. The increases were a signal that the liner shipping industry has slowly adapted to the volume adjustments. By early 2010 the bunker price was already back to USD 450 per ton and hence a lot of measures taken by the shipping lines can be put into question if such increases are to continue.

2.2.3 Implications on market structure The crisis also has an impact on the market structure. A wave of acquisitions and mergers appears inevitable in the medium term, particularly if the recession continues in 2010 and 2011. The drivers for a further consolidation in the liner business relate to the poor financial results of shipping lines which could see some shipping lines default and, to the objective of many shipping lines, to push down costs by increasing the scale of operations. Many leading container shipping lines incurred high losses in 2009 (Table 3.4). In the first half of 2009, liner revenues averaged 20 per cent below operating breakeven. In comparison, Goldman Sachs (2009) reports long-term average EBIT margins (earnings before interest and taxes) of +11.8 per cent in the period 1995-2008. The liner shipping market is cyclical in nature so there have been other periods in which container carriers faced operational losses. What makes the current crisis unique is both the pace and intensity of the present financial problems of most shipping lines bringing a number of them close to bankruptcy. In early October 2009, CMA CGM had to seek a restructuring of a USD 5 billion debt in order to stay afloat (i.e. moratorium on its debt). ZIM Line, CSAV and Hapag-Lloyd have entered into far-reaching restructuring programs and Maersk Line and MSC are also known to be suffering badly.

The crisis increased the diversity among shipping lines regarding long-term strategies. MSC, Evergreen and Hapag Lloyd are among the shipping lines concentrating on the core business of liner shipping. The concept is to invest capital in liner shipping, demanding a return on that capital. While MSC and Evergreen are also present in the terminal business and have some presence in inland logistics,

Table 3.4 Financial Results for a Number of Major Container Shipping Lines

Shipping Line	Operating losses in H1 2009 (Million USD)	Percentage rate shortfall in H1 2009 (*)
Maersk Line	829	8%
China Shipping	475	37%
COSCO	630	38%
Hapag-Lloyd	618	19%
OOCL	197	10%
NOL/APL	379	15%
Hanjin	342	17%
ZIM Line	380	33%

Note: * Shortfall as percentage of average rate (EBIT/Revenue).
Source: Authors' compilation based on Marsoft (2009).

Hapag Lloyd limits itself to operating ships. APL and OOCL reinvented themselves as logistics service providers competing directly against established logistics service providers such as Kuehne & Nagel and DHL. Japanese and Korean lines increasingly rely on their role within large shipping conglomerates. For example, NYK and MOL have only 40 per cent of their business in liner shipping. By being involved in many sectors, these conglomerates spread risk. Finally, the AP Moller group (of which Maersk Line is a subsidiary) and CMA CGM continue to rely heavily on vertical integration with involvements in container shipping, terminal operations and inland logistics. Particularly the AP Moller group has gone beyond container logistics with involvements in supermarkets and the oil business.

2.2.4 Towards a structural shift? Is the reaction of the shipping industry a temporary adjustment or is it the start of structural shifts in the industry? The answer lies in at least three considerations: redesign of service networks including adjustments to ship size and routing patterns, competition among lines and go-forward strategies on capacity deployment, and review of lines' business models.

First of all, the adjustments made are historic in proportion: never have so many vessels been taken out of service or have shipping lines redesigned their liner service networks. There is a common belief that the market will not see an increase in the maximum size of container vessel for at least the next five years. The Emma Maersk class and comparable vessel sizes of MSC (cf. the MSC Beatrice of around 14,000 TEU) are thus expected to form the upper limit in vessel size, at least for the coming years. The crisis has also urged shipping lines to rationalize services and to cascade larger vessels downstream to secondary trade routes. Low charter rates might give incentive to outsiders, such as large logistics groups and financial groups, to consider entering the liner shipping market. Routing patterns are being reassessed. For example, the crisis led to the rediscovery of the Cape route due to a combination of high Suez Canal fees, low vessel loading factors and high insurance fees caused by piracy problems near Somalia. The worldwide shifts in cargo flows (e.g. strong growth of flows between South America and the Far East and Africa and the Far East) combined with an active hub port strategy of South Africa (Notteboom 2010), Mauritius and others have given a strong opportunity for ports in Sub-Saharan Africa to take up a more prominent role in the world shipping networks.

Secondly, the abolition of liner conferences in Europe since October 2008 has led to an interesting competitive game with respect to capacity deployment. Some shipping lines see the crisis as an opportunity to gain market share. MSC, the world's second-largest ocean carrier, is closing the gap on market leader Maersk Line, by rapidly expanding its fleet during the slump in container shipping, while its rivals shrink their capacity. This diverging strategy in a first phase paid off by increasing its share in container volumes. In a second phase, it clearly negatively affected the financial base of MSC rapidly emptying its deep financial pockets. The lack of solidarity and coordination among shipping lines when it comes to capacity deployment potentially leads to a trigger point where the fragile rate restoration process might be undermined by shipping lines wanting to reverse their strategy. Capacity management thus proves to be a very difficult issue in shrinking markets, as the lines which decide to cut capacity might see other shipping lines "free riding" on the resulting rate restoration.

As a third consideration in assessing whether a structural shift is occurring, the crisis is a good opportunity for shipping lines to make a comprehensive review of their business models. The pace and timing of economic recovery will be an important factor to shipping lines. The containership sector will likely continue adjusting rates up to a level where carriers may achieve mid-cycle margins and returns by 2011-2012. Demand is expected to grow at a slower pace than previous up-cycles, while deferred deliveries and idled fleet are likely to cap upside to rates and returns until supply is fully absorbed, which could take four to five years from 2010. It is evident that any path to recovery will have to cope with structural economic factors such as the need for covering the accumulated debt, either by paying it back or through default, and the impact of higher unemployment on consumption patterns. A path of slow recovery would allow shipping lines to redeploy resources and equipment step by step without destabilizing the market and the rates. However, a steep recovery would take everybody by surprise and would put strong pressure on making idle capacity available to the market in a short period of time. Such a fast development towards high demand could lead to a situation of (infrastructural) capacity constraints, soaring rates and soaring fuel prices. The bigger companies are also likely to expand their control of the market, to reassess vertical integration strategies in the chain, and take over slots and terminals from smaller competitors.

How shipping lines actually respond to economic the downturn and its aftermath under the three factors posed for consideration remains to be seen, but it seems likely that the response will be more wholesale than piecemeal, more a permanent adjustment than a temporary band aid.

2.3 The Dynamics Driving the Terminal Operating Business

Whereas a decade ago the container handling sector was still rather fragmented and characterized by about ten large players, the picture looks drastically different today. The worldwide container handling industry is nowadays dominated by four worldwide operating companies (PSA, HPH, DP World and APM Terminals), representing some 42 per cent of total worldwide container handling. In developing a global expansion strategy the four dominant operators try to keep a competitive edge by building barriers to prevent competitors from entering their domains or against them succeeding if they do. These barriers are partly based on the building of strongholds in selected ports around the world and on advanced know how on the construction and management of container terminals. It is increasingly difficult for new entrants to challenge the top global terminal operating companies.

Global terminal operators are increasingly hedging the risks by setting up dedicated terminal joint ventures with shipping lines. As terminal operators are

Table 3.5 Major Port Terminal Acquisitions Since 2005

Date	Transaction	Price paid for transaction compared to EBITDA
2005	DP World takes over CSX World Terminals	14 times
Early 2006	PSA acquires a 20 per cent stake in HPH	17 times
Mid 2006	DP World acquires P&O Ports	19 times
Mid 2006	Goldman Sachs Consortium acquires ABP	14.5 times
End 2006	AIG acquires P&O Ports North America	24 times
Early 2007	Ontario Teachers' Pension Fund acquires OOIL Terminals	23.5 times
Mid 2007	RREEF acquires Maher Terminals	25 times

Note: EBITDA = Earnings Before Interest, Taxes, Depreciation and Amortization.
Source: Authors' compilation.

urged towards a better integration of terminals in supply chains and shipping lines are acquiring container terminal assets worldwide, leading terminal operating companies are developing diverging strategies towards the control of larger parts of the supply chain. The door-to-door philosophy has transformed a number of terminal operators into logistics organizations and/or organizers/operators of inland services. Prior to the crisis, the scarcity of land for terminal development (particularly in developed economies), excellent prospects for container growth and high returns on investment (in many cases 15 per cent or more) attracted many investors. More and more financial suitors such as banks, hedge funds, private equity groups and investors entered the terminal business in the period between 2000 and 2007 (Babcock and Brown, Macquarie Infrastructure and American International Group to name a few). Global terminal operators and investor groups paid record prices for port assets (Table 3.5).

An important driver in the acquisition of port terminals by the financial actor is the assumption of liquidity, implying that it is possible to rapidly sell terminal assets if needs be as a buyer could readily be found. In a market where containerized flows are growing, terminals are indeed fairly liquid assets, but they can quickly turn to be illiquid if market conditions change. In retrospective, higher terminal asset prices beget even higher prices, which can be seen as a port terminal bubble with its resulting overinvestment. This would represent a dramatic shift from an enduring perception of lack of capacity towards a belief in overcapacity.

2.3.1 Reaction of the terminal operating industry to the downturn Terminal operating companies have to a certain extent adjusted their strategy to cope with the drop in container volumes since late 2008. Vertical integration strategies of terminal operating companies have been stalled. For instance, until early 2008 PSA Europe was actively looking into the development of a hinterland strategy in Europe. With the crisis, all attention is back on the seaport terminal operations. Also other major terminal operating groups such as HPH and DP World are revising their hinterland strategies which could eventually lead to a (temporary?) reversal of their direct involvement in barge services, rail services and inland terminals.

During the peak years which preceded the economic crisis, accountants, venture capitalists, financial speculators and pension funds with no or little knowledge of the terminal operating business assumed increased importance in the global terminal operating companies, and shipping lines were not sure whether they should establish a specialized terminal handling company, get into a joint venture with either an terminal operator or another carrier, or both. Besides, governments and port authorities started to become quite greedy when tendering for the operations of their port facilities (generally container terminals). As a result extremely high prices were paid for facilities that at best were second hand (see Table 3.5 earlier). Due diligence became a formality and expected net returns on investment and project IRRs (internal rate of return) were grossly overrated as they were based on the belief that container throughput figures would continue to rise, container handling facilities would be in short supply and hence prices would rise steeply. A similar reasoning was made by bidders for new terminal developments whereby the container facilities were expected to be full almost immediately after their commissioning and handling tariffs would follow an ever rising curve. To make sure that the project would not escape them, bidders (even the more experienced and rational) were ready to put in bids that far exceeded the conditions of a reasonable offer. Not only were they committing themselves to huge investments and tariff reductions, but they were also accepting excessive risks.

The crisis led to a sudden decline in the attractiveness of terminals as a result of existing cash problems among many companies and a fear of structural overcapacity in the market. The current overcapacity forces down tariffs and thus undermines the ROI. Currently, most terminals are frantically looking for additional clients, ships and cargo. The argument that port throughput and in particular the container throughput will come back to acceptable levels overlooks the fact that a serious consolidation on the supply side of shipping is inevitable and that many carriers may leave container shipping altogether, or at least specific segments of the market. The changed economic situation means that terminal operators have adopted a more cautious assessment of future prospects. We observe a clear slowdown in investments from global operators, shipping lines and financial institutions in container ports globally. In April 2009, DP World announced that more than half of their terminal plans are (seriously) delayed and some are even canceled. At the same time, the problems facing DP World in 2009 were also clearly linked to major financial problems at its mother company, Dubai World, and especially those at its sister company, Nakheel, one of the main real estate developers in Dubai.

The general economic slowdown may well result in some investors having to sell terminal interests and this may create opportunities for those global terminal operators and financial investors with ready access to the necessary funds. The most likely terminal portfolios which may become available are those owned by shipping lines. This does not preclude genuine terminal operators' portfolios, particularly if they have over invested in new or expanded terminal assets.

Of key interest in any M&A activity is the valuation of port and terminal assets. In the peak period of demand growth and interest in acquiring terminals during 2005-2007, port companies were being valued (and paid for) at EBITDA multiples in excess of 20 times. Anecdotal evidence suggests that multiples of around 8-12 times EBITDA are the new benchmark, but there has yet to be any major M&A deals go through to verify these new levels in the market.

The financing of large terminal projects has become a more difficult task than before. This picture is quite a shift from a few years ago when financial means for terminal projects were widely available and when the scarcity of port land for further expansion constituted the major concern to terminal operators. However, partly because of the scarcity of land, ports should be considered as long term investments and for those cash rich, now may be an opportunity to take advantage of the low prices forced on those selling their assets.

3. Second and Third Ramifications

The analysis so far has focused on the ramifications directly related to the maritime shipping industry, particularly shipping lines and port operators. The sudden adjustments observed will also have profound ramifications on the added value function of global supply chains as well as the level of concentration of cargo flows. These potential changes are more medium and long term in scope, implying that they will be dealt more from an assumption standpoint.

3.1 Changes in the Added Value Process: From Capture to Retention

The generation of value in a supply chain is a process that mainly takes three major forms (Henderson *et al.* 2002):

- *Value creation or capture*. Concerns entirely new activities within a supply chain and is linked with a paradigm shift such as a new terminal, lower distribution costs, a new technology, a new market, etc. This process involves the higher return in terms of added value since new activities are created.
- *Value expansion*. The growth of existing strengths, mainly in relation with the growth of traffic along a supply chain.
- *Value retention*. Keeping desirable added-value activities which under existing circumstances would have ceased and/or relocated elsewhere. It is a difficult process to mitigate since it is linked with changes in economic fundamentals.

Under the current context, it is becoming clear that the emphasis for many transport chains will shift from the creation, capture or expansion of added value to strategies aimed at retaining what has been gained in light of growing competition and the lack of pricing power. One of the first advantages to be impacted by a recession is the pricing power of key players, forcing a spatially and functionally unequal rationalization of supply chains. Intermediary locations that were able to extract value from freight flows in a context of enduring growth see their business model being compromised by a debasement of the process. As seen before, maritime shipping has curtailed much capacity, even if in hindsight it was at a rate that was slower than the real drop in demand, sharply impacting freight rates and their capacity to extract revenue. Port operators are in an even more complex situation since their capacity cannot be changed (reduced) effectively and they have limited margin to expand their hinterlands. Like maritime shipping companies, their pricing power is reduced and several will struggle to retain their added value activities.

The terminal overcapacity situation has made terminal operators less stringent on dwell times with cargo able to remain stored at terminal facilities for longer periods of time without penalty. An additional level of terminalization of supply chains could increase the time and cost flexibility of supply chains (Rodrigue and Notteboom 2009a). The dynamics of the maritime/land interface are impacted with a rebalancing of the share of the gateways versus the hinterland in the value capture process, and this is to the advantage of the hinterland. Pressures on leasing rates and demurrage can also be felt as maritime shipping and leasing companies see lower utilization levels of their containerized assets. Leasing terms are becoming more lenient with the resulting improvement in the flexibility of empty container repositioning. An important added value activity that could be impacted by such a process is transloading. The rationale of transloading where the contents of maritime containers are transloaded into domestic trucks for demurrage could be increasingly by-passed, leading to a loss of an added value activity at gateways but additional opportunities may arise inland with a better availability of maritime containers, at least if the lines are ready to make more containers available in the hinterland.

The debasement of the value capture process of several containerized supply chains would thus incite the search for cargo, which would benefit inland locations. As shippers ponder the rationalization of their supply chains, value capture could further move inland particularly at locations able to provide return cargo instead of just empty containers to be repositioned. The selection of transport chains would be more a factor of total return for the shippers than specifically servicing the needs of their customers in terms of frequency and time. This could also lead to additional strategies aiming towards the containerization of commodities as a way to better anchor cargo, which could help redirect container flows towards gateways, corridors and inland ports able to provide more balanced trade options.

3.2 Spatial (De)concentration of Cargo Flows

At this point, the rate of decline in port traffic does not appear to be related to size, namely among the world's largest ports. A reconfiguration may be observed first as maritime shipping companies restructure their networks according to demand patterns, leading to traffic changes (mostly negative) for the concerned ports. However, as further rationalization takes place, shipping companies face a more comprehensive review of their port calls and network configurations. Port pricing plays an important role in this reconfiguration with the larger ports and their more developed hinterland transport systems in better position than the small and medium-sized ports.

There are signs that the current drop in volumes leads to an increased geographical specialization of gateway ports vis-à-vis specific overseas maritime regions. For example, shipping lines have started to consolidate most of their vessel calls on the Far East – North Europe trade in Rotterdam. Consequently, small ports (e.g. Amsterdam) and even large ports (such as Hamburg and Antwerp) in the region have lost Asia-related calls. However, at the same time Antwerp has succeeded in further reinforcing its strong position on liner services to Africa, the Middle East, India and the Americas.

Regarding this, a remarkable dichotomy could emerge between ports and their hinterland because of changes in cargo flows. The hinterland is more flexible to handle volume changes since it is less subject to economies of scale. It can deal with a significant scale back while being able to maintain a similar level of performance, particularly for trucking which is close to being indifferent to traffic changes on a unit basis. For ports, the volume handled is a strong factor in the productivity and competitiveness of terminals. There is a limited capability, particularly for terminal operators, to handle the financial ramifications of a drop in cargo volume, particularly since terminals are financially leveraged with enduring growth expectations. This generates a new dynamic equilibrium between maritime and inland transport systems. Ports that have a relatively captive hinterland in proximity have more latitude than ports that have a more long distance and fragmented hinterland or depend heavily on transshipment. Also, ports that have a more balanced traffic are likely to be less impacted. It is, however, not entirely clear at this point to what extent sustained declines in traffic could be a factor of cargo concentration or deconcentration.

The pricing power remains a powerful factor with shipping companies even more sensitive to cost considerations in periods of economic downturn. Still, this trend may not always favor the concentration of cargo flows. For instance, the ports of Los Angeles and Long Beach decided to be more environmentally stringent at the wrong time, imposing various fees and restrictions on gate access. An outcome is that maritime shipping lines are considering moving cargo from the largest west coast hubs to other ports along the range. MSC would offer more calls to Vancouver with Maersk and CMA-CGM having more calls to Seattle. By offering a behavior that is perceived as less rent seeking, several smaller ports try to mitigate traffic pressures since many may not have been heavily involved in capital intensive expansion projects, which in a context of lower traffic expectations undermines competitiveness. The need to provide minimum facilities to serve the lines will in any event involve ports that have at least some additional capacity to handle the bigger ships or they will not play any substantial role.

Finally, it is also important to mention that concentration and de-concentration patterns will also be influenced by the responses of governments, although it is beyond the scope of this chapter to fully address this additional source of uncertainty. While the economic crisis has slowed down infrastructure investment in some places, other ports are being expanded according to plan and with government support (e.g., Maasvlakte 2 in Rotterdam is being constructed following a time schedule set before the crisis).

4. Conclusions

In this chapter we have provided an analysis of what we identified as three major ramifications of the current crisis on the maritime and logistics industry.

The first and dominant ramification dealt with the pricing and investment strategy of shipping lines and terminal operators and the dichotomy between integration and disintegration within value chains. Based on the analysis provided, we argue that the current crisis has led to individual capacity strategies in the liner shipping industry that do not necessarily converge. This observation is a potential source for instability in the market, as free riders seek short-term gains on the back of other shipping lines which are idling capacity. A sound industry-wide capacity management is a prerequisite for avoiding a path of destructive rate competition which could form the base of a hefty period of peaks and lows in the liner industry, even beyond the current economic crisis, but such a capacity management system has also to be in line with the competition policy of national and supranational governments. In such a market environment, a vertical integration strategy could help shipping lines to decrease their strong reliance on an unstable liner shipping business.

Secondly, we argue that temporary measures of shipping lines will likely have long-term effects on the routing of goods. The emergence of the Cape route could be an example of such a development path. The crisis thus opens windows of opportunity as it forces market players to search for new, sometimes uncommon, solutions.

Thirdly, the crisis might have far-reaching effects on the (financial) attractiveness and capacity situation in the container terminal business. While the long-term prospects might not be as bleak since land scarcity will remain a major concern, the industry in the short and medium terms will have to deal with overcapacity and lower returns.

For the two other ramifications, only preliminary assessments can be made so far. The second ramification involves how the forces behind the added value creation process are reconciled with the debasement of value chains through competition and efficiency improvements. While the value capture process is a much sought after effect, particularly by gateways between major systems of circulation, evidence underlines that in light of uncertain future traffic flows, many elements of the transport chain will switch towards value retention strategies. This will likely involve more flexibility in the usage of containerized assets, such as initially longer dwell times, and permit a rebalancing of the respective share of added value activities taking place around gateways versus inland locations. This is possibly to the advantage of the hinterland. The most important driver of value capture and expansion would be linked at setting more balanced transport chains with return cargo with attempts at better integrating some bulk within containerized supply chains a fundamental part of such a strategy. For many inland ports, this would become a crucial factor to insure their viability on the medium term.

The third ramification pertains to the duality between concentration versus deconcentration, particularly in terms of ports, gateways and corridors. The last decade has been the object of a process of traffic concentration, but this concentration consistently involved opportunities for smaller ports and inland terminals to capture additional traffic as containerization diffused geographically and functionally. The rebalancing of freight flows either towards concentration or deconcentration remains in flux. While large ports, because of their pricing power and better hinterland access, could be less impacted than smaller ports by a rationalization of network configurations, rent seeking behavior imposed by prior infrastructure investments could advantage players that are less financially leveraged. Again, this raises the question to what extent financial considerations, as opposed to macroeconomic considerations, could be an explanatory factor in the traffic performance of transport chains, including maritime routes, gateways and inland ports.

All this put together, particularly in consideration to the across the board container traffic decline, we argue that the evidence supports more the paradigm shift thesis than a temporary correction. A thorough and more structured analysis of the full ramifications of the crisis on the container industry structure will only be possible years from now once the market has adjusted to a new financial and macroeconomic context. Here, we have had to limit ourselves to the identification of key trends, developments and potential impacts.

Chapter 4
Carriers' Role in Opening Gateways: Experiences from Major Port Regions

Antoine Frémont and Francesco Parola

1. Introduction

Over the last decades, the development of containerization and intermodalism has deeply changed the port-hinterland relationship (Van Klink and Van der Berg 1998). The enhancement of inland corridors contributed to a paradigm shift from *captive* hinterlands to *contestable* hinterlands. At the same time, container transport chains experienced a profound institutional turn in port and intermodal operations. The port reform process at the international level in both advanced and developing countries brought to the fore the port-as-landlord model allowing the entrance of private players to the terminal activities and, in some cases, leading to the construction of vast facility networks managed by international corporations.

More recently, the dramatic throughput growth worldwide progressively led to the emergence of congestion in some major "historical" ports such as Hong Kong, Los Angeles, Hamburg, and Rotterdam. The need of additional handling capacity often drove port operations to escape from the old city sites toward new deepwater sites in the metropolitan surroundings and in what Bird (1963) described as "outports". This process is called port regionalization, a trend characterized by the rise of centrifugal forces causing the fragmentation of port activities (Notteboom and Rodrigue 2005) as well as the de-concentration of container throughput within some ranges (Slack and Wang 2004).

The above trends suggest the application of the lifecycle theory in ports (Sletmo 1999). This theory, developed by Vernon (1966), was originally applied to describing product evolution and then extended to world markets (Dicken 1998). Looking at traffic trends in port ranges, some analogies with industrial markets clearly emerge. There are four stages in this model. In the introduction (i), some ports in a range start to convert their facilities to the container business and to lease terminals to private terminal operators. Traffic flows are still low and scattered in a few ports. In the growth phase (ii), container volumes increase at high-growth rates. A much larger number of ports is involved in the development of the range but the throughput is concentrated in a handful of selected gateways (load centres). Both pure stevedores and integrated carriers increase their engagement in port handling operations. In the maturity stage (iii), the port space becomes scarce and the congestion of the system increases. The saturation of the major ports leads to the initial development of new "satellite" ports in the surroundings. Traffic flow concentration within the range may decrease. In the (eventual) declining stage (iv), the market share of the range decreases in respect to neighboring ranges. Some private actors disinvest and transfer their operations to other regions.

Within this framework, ocean carriers take on a leading role as they can affect the success of a port and/or a region. Carriers define port hierarchy in various ranges and select gateways lying in different lifecycle stages. Academic literature achieved a good understanding of the above macro trends, outlining development patterns in major regions and identifying the drivers of such changes. In this respect, many evolutionary models were proposed and discussed (Bird 1963; Taaffe *et al.* 1963; Hayuth 1981; Notteboom 2006b). However, very fragmented and still exploratory contributions have been made on the role of shipowning firms in these transformations (Notteboom 2004b; Lam *et al.* 2005). This chapter is a contribution to this lack of analysis.

Our main hypothesis is that on each maritime range, each shipping line tries to maintain at least one main port where it can concentrate the calls for its fleet. This main port acts as a hub in the shipping line's maritime network and can be considered as the gateway for this carrier on this range. The concentration of calls at this gateway does not exclude the use of other ports in the range, but these other ports have a secondary role with regards to this main port. The gateway port for one company does not need be the most important port on the range for all shipping lines, but for that one company it takes on this role. This chapter, therefore, investigates the role played by leading shipping lines in reshaping inter-port competition through the selection of their own gateways.

2. Shipping Lines' Concentration in Ports

In this section, we shall consider the case of port ranges that serve a hinterland. The ports in the maritime range in question are thus primarily hinterland ports, even though they may also perform transhipment operations. The concentration of shipping lines in ports is the process by which a shipping line concentrates its maritime and terrestrial activities in a certain port in a maritime range. As a result of this concentration, it achieves a dominant position in the port, which provides it with certain advantages (Frémont and Soppé 2007).

2.1 The Factors Which Encourage Shipping Line Concentration in Ports

The principal factors that encourage the shipping lines to concentrate their calls at a port are essentially related to the increasing size of vessels. The greater a vessel's capacity, the less justified it becomes to make calls involving only a small number of container movements. The reason for this is that the operating costs of a vessel increase with its size, which automatically favours a reduction in the number of calls (Ashar 1999; Cullinane and Khanna 2000; Stopford 2009). Based on a number of interviews with representatives of shipping lines, at least 10 per cent of a vessel's capacity must be handled to justify a call. In theory, the best turnaround time for a vessel would involve just two ports, with unloading followed by full reloading in each of them. This perfect situation for shipping lines would require the availability of adequate volumes for transport in both ports and balanced traffic. Concentrating calls for several principal routes in a single hinterland port opens up certain possibilities for the shipping line both with regard to its clientele and its operating policy, particularly with regard to container logistics. As far as its clients are concerned, the shipping line will be able to broaden its commercial offer, providing a larger number of maritime services, and therefore more geographical destinations. Logically, it will also tend to offer higher frequencies, as more than one maritime route may serve a given port. With regard to container logistics (Frémont 2009a), the shipping line can more easily reposition its empty containers in view of the imbalances between the different markets.

This concentration also opens up the possibility of transhipment between mother vessels in order to extend the commercial offer and improve container logistics, or between mother vessels and feeder vessels to serve secondary markets. The principal port thus ceases to be simply a hinterland port for the shipping line but performs the combined functions of a hinterland and a transhipment port.

New possibilities are also opened up for the shipping line to provide hinterland services. The concentration of calls means that large volumes are required to fill the vessels. Unless the port has a very rich near hinterland, it is necessary to collect the freight in question some way inland, which necessitates the use of consolidated modes such as rail or inland waterways. Consolidated transport provides a way of collecting freight at the lowest possible cost, which makes it possible to offer lower prices than the competition and limit the cost for the shipping line. The

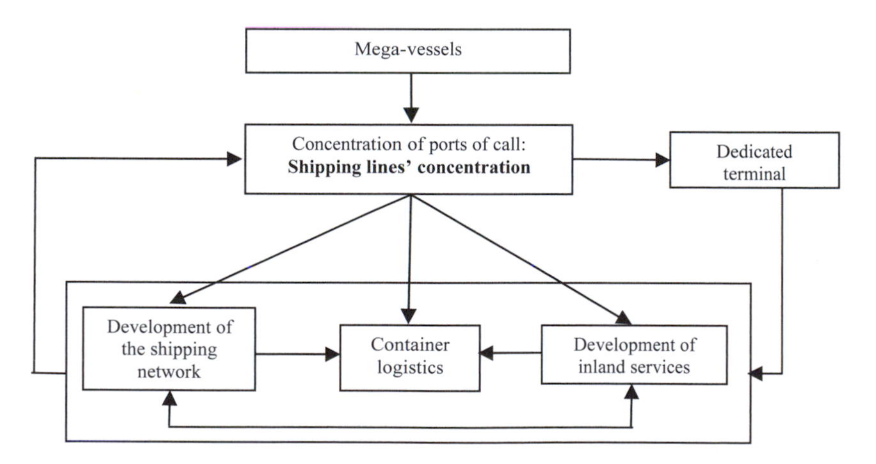

Figure 4.1 Shipping Line Concentration
Source: Authors.

shipping line can set up its own inland services, often by chartering trains or barges from specialized operators. The shipping line can organize carrier haulage with its largest clients in order to control the entire transport chain and this also helps to improve container logistics (Parola *et al.* 2006).

To set up a port concentration policy of this type, the shipping line must have a dedicated terminal, whose size will depend on the scale of its activities. This dedicated terminal is an absolute necessity as, in particular, it provides a means of securing the vessel's calls without depending on competing shipping lines (Harrigan 1984; Levy 1985; Stuckey and White 1993). One possibility that is available to the shipping line is to own a dedicated terminal but use the services of a specialized freight handler. But, in most cases, it is necessary for a shipping line to invest in a dedicated terminal. The shipping line may either have a minority or a majority share in the capital. When the shipping line participates in the capital it means that its activities must be long-term, as concessions often run for 25 or 30 years. This long-term commitment strengthens the process of shipping line concentration in ports, particularly with regard to terrestrial activities. This is because to set up inland rail or waterway services, it is necessary to build customer loyalty among the shippers who have the potential of providing large volumes of traffic over a long period of time.

In this way, there is a snowball effect by which the commercial offer, the development of inland services and container logistics strengthen the effect of shipping line port concentration and the need for a dedicated terminal. The principal port thus becomes the shipping line's gateway port in the maritime range.

2.2 What are the Advantages of a Dominant Position?

In a port of this type, it is in the shipping line's interests to have a dominant position (Abell and Hammond 1979), that is to say one in which it is the favoured and indispensable discussion partner for the port administration due to its ownership of one or more dedicated terminals. It is difficult to fix a precise threshold beyond which a shipping line can be said to occupy a dominant position. In a very large port, we consider that this dominance is acquired when the shipping line handles more than 15-20 per cent of the port traffic, while in a secondary port the share can easily reach or exceed 50 per cent because of weaker competition.

A shipping line may wish to attain a dominant position for two main reasons. It does not wish the expansion of its activities to be hampered, slowed down or disrupted by the other players present in the port, whether these are direct competitors or potential partners in the organization of the transport chain. In addition, a dominant position provides it with considerable negotiating power in relation to these various partners (Porter 1985; Grant 2010).

A dominant position means that the shipping line in question automatically becomes a favoured client for the port authority. This makes it easier for the shipping line to negotiate port tariffs, and means that it is in a position of strength if the port authority proposes a new terminal concession and even means it may cooperate to some extent with the port authority in the design of terminal facilities or inland services from the terminals.

With regard to freight handling, since the shipping line generates very high volumes of traffic, it may perform its own freight handling, be the favoured or indeed exclusive customer of a freight handling company or develop hybrid forms of cooperation with one.

The shipping line will be in a similar position of strength in relation to inland transport operators: road hauliers, rail companies or inland shipping lines. By providing very large long-term volumes, the shipping line will be able to charter trains or barges at low prices and guarantee regular frequencies (Franc and Van der Horst 2008).

In a port of this type, competing shipping lines have only a marginal role and are not in a strong negotiating position. On the face of it, their presence at the port in question provides no benefits. It is not at all certain that it is in the interest of freight forwarders to increase their presence in a port of this type either, as a dominant shipping line is primarily interested in setting up carrier haulage which does not provide a market for freight forwarders. A port of this type that is dominated by a shipping line is therefore first and foremost positioned in a corridor that operates on a door-to-door basis, unless the shipping line itself develops logistical services for the freight, usually via a logistical subsidiary.

In spite of the possible benefits of the policy of shipping line concentration at ports, the shipping lines do not necessarily manage to implement it. This is because doing so entails concentrating a large number of routes in a single port while investing in a dedicated terminal or inland services. Each shipping line has to make trade-offs, mainly with regard to the nature of its maritime network and/or its financial constraints. The policy is therefore implemented in an uneven or limited manner, for example in one or two of the world's maritime ranges.

With such dominating positions, the shipping lines are well positioned to become the privileged actor in negotiations with port authorities, and are able to use their influence in favour of their expansion projects or of hinterland servicing. Thus, these shipping lines seek, whenever possible, their own gateways. To the contrary, in the more general gateways such as Hong Kong and Shanghai, no shipping lines gets to hold a dominating position.

3. Methodological Notes

3.1 The Selection of Maritime Ranges

This study focuses on the world's major gateway port regions. With the exception of West Africa, maritime ranges representing less than 2 per cent of the world's traffic have been excluded from the analysis. Ranges where the main port's activity is mostly transhipment are also not included. This provision eliminates South East Asia, the Middle East, and Central America from the analysis. In our

Table 4.1 Market Share of the Selected Ranges

Regions	Ranges	Throughput share (%)			Annual growth 1994-2006 (%)
		1994	2002	2006	
East and North East Asia	East Asia	30.5	38.9	44.6	13.8
	North East Asia	18.1	14.6	12.3	6.8
North America	East Coast	10.4	8.5	6.9	6.5
	West Coast	12.9	11.1	10.0	8.0
Europe mainland	Northern Europe	16.7	14.8	14.0	8.7
	Southern Europe	5.4	5.6	4.7	8.9
Other ranges	West Africa	1.1	1.1	1.0	9.6
	Turkey and Black Sea	1.6	2.0	2.5	14.6
	South America East Coast	3.2	3.4	4.1	12.6
Total throughput (million TEUs)		76.0	153.4	246.3	10.3

Source: Authors' calculation from data available on www.ci-online.co.uk.

sample, we have selected the most relevant areas in terms of gateway traffic: North America both East and West Coast, East Asia, North East Asia, Northern Europe and Southern Europe. Moreover, two emerging ranges have been included in the sample: South America East Coast, Turkey and the Black Sea. West Africa is also in the sample as an example of a range depending on an underdeveloped region (Table 4.1).

Our sample represents approximately 70 per cent of the world's gateway traffic (2006). The composition of the ranges has been derived from *Containerisation International*.

3.2 Shipping Lines Behaviour Analysis

We analysed the Weekly Containerised Transport Capacity (WCTC) deployed by major carriers in the selected ports (Soppé *et al*. 2009). Using this database, we collected information on the major 26 shipping lines. For the specific purposes of this study we isolated data on a sample of carriers, i.e. leading independent players (Maersk Line, MSC, CMA-CGM and Evergreen) and major global alliances (Grand Alliance, CKYH, The New World Alliance). Moreover, in order to evaluate carriers' vertical strategies in ports, we have also collected equity shares information on the investments made by shipping lines in port facilities.

The analysis of carriers' behaviour in selecting and opening new gateways was performed comparing data of various years: 1994, 2002 and 2006. This permits an appraisal to be made of the recent effects (since mid-1990s) produced by reforms in ports and intermodalism. The strategies of the selected shipping lines were then compared with the overall market trends (i.e. the 26 carriers' sample).

The major sources we used are the *Containerisation International* on-line database (for the port throughput and the WCTC) and Drewry Shipping Consultants (for shipping lines' investments in terminals). Some additional information was obtained from port authorities and carriers' websites.

Our analysis is qualitative. Starting from empirical data we make crossed aggregation per port, range and carrier, in order to open a discussion on gateway ranges' development and carriers' strategies. More specifically, for better appreciating trends in the sampled ranges, we apply a measure of inequality, the Gini index (Allison 1978) already introduced in maritime studies (McCalla 1999a; Notteboom 2006b). There are various formulas of this index. In our case, we use the normalized index ($0 \leq G' \leq 1$), for a range composed of n container ports (observations). First, we apply the Gini formula, coinciding with the area between the Lorenz curve and the diagonal of equal distribution:

$$G = 1 - \sum_{k=1}^{n} (X_k - X_k -)(Y_k + Y_k - 1)$$

where X_k is the cumulative percentages of number of ports and Y_k is the cumulative percentages of container throughput. Then, we calculate the maximum Gini value, given by:

$$G_{max} = \frac{n-1}{2n}$$

Finally, their ratio gives the normalized Gini:

$$G' = \frac{G}{G\ max}$$

If G' = 0 there is equidistribution, while if G' = 1 we have the maximum concentration (i.e. inequality).

4. Shipping Line Concentration in Ports: Different Constraints According to the Type of Maritime Range

It is not possible to implement a policy of shipping line concentration in ports in a uniform manner in all port ranges, because the constraints differ according to the type of port range. Some offer a more favourable context than others for this process. For example, quite logically, it is more difficult for a shipping line to achieve a dominant position in a port which is at a high level in the global port hierarchy, where all the shipping lines are present, than to achieve such a position in a secondary port. Four types of maritime range can be identified, which do not provide the same opportunities to shipping lines for implementing a shipping line of port concentration policy.

The only ports considered here are those whose throughput was higher than 400,000 TEUs in 2006. As a consequence, 103 ports on nine ranges were studied.

Based on the statistical analyses through the Gini coefficient, different kinds of evolution can be observed (Table 4.2), partially derived from the Vernon typology (1966).

4.1 Type 1: Very Young Ranges

This first type consists of very young ranges (introduction stage) taking their early steps in deep-sea services. Leadership and hierarchy are still being established. These maritime ranges belong to developing countries. Uncertainty is the rule with regards to the port hierarchy because of the geographic instability associated with these countries.

The West Africa range belongs to this type of range. In West Africa, two main ports should take on domination positions: Abidjan and Lagos. Both are the economic capitals of the two most important countries of the areas. Still, the long standing political instability in these two countries no doubt accounts for part of the loss of these two ports and the search for new gateways by operators on this range.

Because of the instability of the port hierarchy, Type 1 maritime ranges provide opportunities to shipping lines to become dominant in a port. But it could be a risky gamble due to political instability or corruption. Furthermore, the competition being weaker on these secondary ranges, a dominant position in a port is not as necessary as in ranges where the competition is fierce.

4.2 Type 2: Fast Growing Ranges

The second type consists of fast growing ranges, where new ports appear, and where the port hierarchy is changing over time. These fast growing ranges are facing a reorganization of the overall hierarchy. They have not yet reached maturity. They depend on the hinterlands of emerging countries with very dynamic economies and very high annual GDP growth rates.

The East Asia range is the most perfect example of a Type 2 range with the very strong growth of the Chinese ports. In the year 2006, 21 ports had a throughput over 400,000 TEUs. Of these, 15 had a very low to non-existent throughput in 1980. Some, such as Shanghai or Shenzhen, have risen to the top of the hierarchy, challenging Hong Kong's absolute domination and forcing ports like Kaohsiung or Manila even further down the list.

Also belonging to this type of range are Turkey and Black Sea and South America East Coast ranges even though the port hierarchy has not changed as much as on the East Asia range. Not taking into account transhipment ports, newer ports have appeared on these two maritime ranges such as Ambarli and Izmir in Turkey or Rio Grande and Itajai in Brazil.

These maritime ranges are thus facing instability because of the spectacular economic development of their respective countries: China, Turkey or Brazil. On these ranges, shipping lines must cover widening geographic markets in order to answer to the needs of freight forwarders or shippers. Meanwhile, the emergence of new ports gives shipping lines more latitude to operate different port choices, specifically with regards to our hypothesis, to get established in fast growing ports where they can achieve a dominating position compared to their competitors.

4.3 Type 3: Ranges Approaching Maturity

The third range type consists of rather mature ranges where major ports are still consolidating their leaderships. These ranges are characterized by a strengthening of the main gateway. Contrary to the Type 2, Type 3 ranges are characterized by a very slow change to the port hierarchy as well as a reinforcement of the dominant pole. In our sample only one range fits this description, the North American West Coast. The gateway formed by the tandem Los Angeles/Long Beach is largely strengthening its dominant position to the loss of other ports except Vancouver.

In this range, Los Angeles and Long Beach are the two dominant ports that have to be served. For shipping lines, they are therefore the major gateways not only to the close hinterland formed by the huge metropolitan area of Los Angeles but also to the distant hinterland through rail services across the continent. Therefore, no carrier can be in a dominant position. However, the option of safeguarding some terminals through long term concessions in the port is still a valid one.

4.4 Type 4: Ranges Managing Saturation

This fourth type consists of mature ranges, in some cases approaching a declining stage, where the historical and major port(s) are (is) often congested. Also for inland transportation, the main port(s) are (is) losing competitiveness in favour of other ports/ranges.

In the fourth type of range, the traffic growth is less than in Type 2 and 3 ranges. These main ports are situated in countries with well developed and more mature economies than those of emerging countries. The main ports have been in a dominant position since the start of the 1980s. This position seems almost absolute and seems to remain unchallenged.

Yet, during the 1980s and the 1990s, the domination by main ports was challenged, not by the creation of new ports which was the case in Type 2, but by existing ports that were gaining market shares and challenging the dominant ports' position. Competition between ports is less open than in Type 2 ranges because shipping lines can only bet on the competition between a few ports. They have fewer opportunities than those in Type 2. Thus, concentration of several services and huge investments in terminals are needed if a company would like to establish a dominant position in a port. More efforts are needed if the targeted port is a major one.

As Table 4.2, indicates Northern Europe, the North American East Coast and North East Asia all are identified as Type 4 ranges. In the first instance, the position of the port of Rotterdam is slowly being challenged by the ports of Hamburg, Antwerp and Bremerhaven. The North American East Coast is much larger geographically than the Northern range. The port of New York has retained

Table 4.2 The Typology of the Maritime Ranges

Typology	Ranges	Consequences for shipping lines
1. Very young ranges, in underdeveloped countries	West Africa	Niche markets Dealing with instability
2. Fast growing ranges facing a reorganisation of the overall hierarchy	East Asia	Covering widening geographic markets.
	Turkey and Black Sea	More latitude to operate different port choices
	South America East Coast	More latitude to establish a dominating position
3. Ranges approaching maturity. Ranges characterized by a strengthening of the main gateway	North America West Coast	Very challenging to be in a domination position Safeguarding terminals through long terms concessions
4. Mature ranges. Ranges managing saturation. The main port is being challenged	Northern Europe	Competition between a few ports
	Southern Europe	
	North America East Coast	
	North East Asia	

Source: Authors.

its very important advantage, but Savannah, Charleston, Virginia and Miami are gaining some market shares. In North East Asia the domination of Japanese ports is challenged by South-Korean ports, mainly Busan, but also by Gwangyang and Incheon. The Southern Europe range also belongs to this type. Three ports could claim the status of main gateway in the early 1980s: Marseille, Livorno and Genoa, but their domination is completely undermined now by the Spanish ports of Valencia and Barcelona.

Shipping lines also have to take into account some geographic markets that are more vigorous than others to explain some of the rebalancing of the port range hierarchy: the Baltic market and the Eastern Europe markets for the Northern Europe range, Spain in the Western Mediterranean and the south east for the North American East Coast.

5. Shipping Lines' Strategies in the Sampled Maritime Ranges

5.1 Port Networks' Diversification

Table 4.3 provides a measure of the Gini coefficient by type of ranges and by type of carriers. Over the three years of analysis, 1994, 2002 and 2006, important

Table 4.3 Port Concentration in the Various Types of Maritime Ranges Per Carrier and Alliance (Gini Coefficient Based on the WCTC)

Type	Ranges	Year	All carriers	Independent carriers				Global Alliances		
				Maersk	MSC	CMA-CGM	Evergreen	CKYH	Grand Alliance	TNWA
1	West Africa	1994	0.41	0.64	-	0.40	-	-	-	-
		2002	0.40	0.41	0.72	0.34	-	-	-	-
		2006	0.28	0.32	0.44	0.32	-	-	-	-
2	East Asia	1994	0.84	0.93	-	0.89	0.91	-	-	-
		2002	0.64	0.80	0.77	0.73	0.76	0.69	0.81	0.84
		2006	0.58	0.68	0.72	0.70	0.71	0.66	0.65	0.75
	South America East Coast	1994	0.49	0.67	-	-	-	-	-	-
		2002	0.39	0.44	0.64	0.88	0.69	-	-	-
		2006	0.35	0.43	0.47	0.47	0.75	-	-	-
	Turkey and Black Sea	1994	0.92	1.00	-	1.00	-	-	-	-
		2002	0.38	0.54	0.49	-	-	-	-	-
		2006	0.36	0.60	0.52	0.68	0.48	0.83	-	-
3	North America West Coast	1994	0.48	0.87	-	-	0.66	-	-	-
		2002	0.47	0.79	1.00	0.86	0.71	0.53	0.62	0.45
		2006	0.47	0.75	0.93	0.62	0.68	0.51	0.54	0.56
4	Northern Europe	1994	0.57	0.67	0.51	0.63	0.63	-	-	-
		2002	0.53	0.66	0.73	0.52	0.55	0.66	0.66	0.66
		2006	0.50	0.61	0.61	0.41	0.61	0.59	0.52	0.59
	North America East Coast	1994	0.50	0.66	0.51	-	0.81	-	-	-
		2002	0.59	0.64	0.67	0.78	0.80	0.76	0.70	0.68
		2006	0.53	0.62	0.49	0.67	0.67	0.68	0.72	0.70
	North East Asia	1994	0.54	0.77	-	1.00	0.71	-	-	-
		2002	0.51	0.67	0.78	0.96	0.53	0.74	0.66	0.57
		2006	0.55	0.66	0.86	0.97	0.57	0.65	0.68	0.68
	Southern Europe	1994	0.53	0.83	0.86	0.77	0.67	-	-	-
		2002	0.51	0.59	0.63	0.74	0.67	0.76	0.83	-
		2006	0.47	0.72	0.60	0.72	0.61	0.67	0.83	-

Source: Authors' calculation from WCTC database.

differences appear with regards to the types of ranges. In Type 2 (ranges in turmoil), the Gini value goes down dramatically. The creation of new ports brings down the port concentration thus favouring the growth of new gateways. In ranges of the second type (where the dominant port is challenged) and the third type (ranges characterized by a strengthening of the main gateway) the Gini value hardly changes. However, for independent carriers as well as for global alliances, the Gini value is slowly getting closer to the average of the sample. Finally, in West Africa, the value is weaker than on other ranges over the timeline. In a developing region, the use of more than one port is a given. Indeed, because of port traffic congestion, weakness of traffic and the lack of hinterland connections, no port can claim to be a gateway for this range.

5.2 The Search for a Dominant Position in a Port: The Case of East Asia

In spite of this general trend, our hypothesis is based on the idea that, as far as possible, a shipping line wishes to establish a dominant position in a port in the sea range in order for it to become its gateway. Since it is more difficult to establish a dominant position in a principal port, some shipping lines will target secondary ports, which are very frequently those where their competitors are not yet firmly established. This is particularly evident when the shipping line is an independent carrier. An independent carrier is more likely to deviate from the generally used network because it does not depend on the will of another carrier. It is easier for it to decide going where its competitors do not go and to differentiate from one another through its particular port network. In contrast, the networks used by alliances are closer to the general network. Independent carriers, however, can differentiate themselves in port choices only if the range allows for that possibility. Maritime ranges belonging to Type 1 are more suitable to this search for differentiation because of the presence of new ports.

The East Asia range is a good example. In 1994, Hong Kong was, without contest, the dominant port for all shipping lines. The only other potential choice was Kaohsiung. In 2002, then in 2006, the situation changed dramatically. All the shipping lines are serving a larger number of ports which explains the decrease of the concentration coefficient and translates into a systematic drop in Hong Kong's role as a main gateway. Individual carrier port selections have thus become more diverse and differentiated (Slack 1985; Malaga and Sammons 2008).

It is possible to differentiate between "traditional" and "independent" carriers. Traditional shipping lines are situated in the heart of the market: they concentrate most of their capacity in the Pearl River Delta, in Hong Kong but also on the port of Yantian in the special economic zone of Shenzhen. They then concentrate, more so than independent carriers, their capacity on the port of Shanghai. Maersk and the alliances are of this configuration as well as Evergreen which, because of its roots, favours Kaohsiung, but less than in 1994.

Two shipping lines have developed alternate strategies. MSC and CMA-CGM concentrate their activities on ports which are not the most important in terms

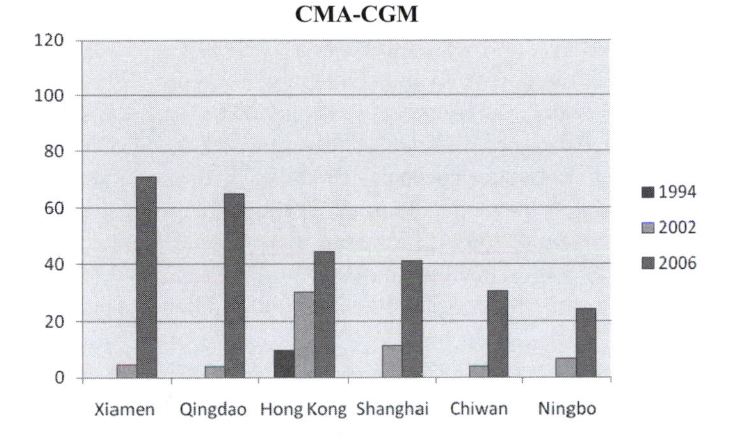

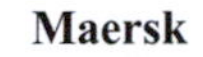

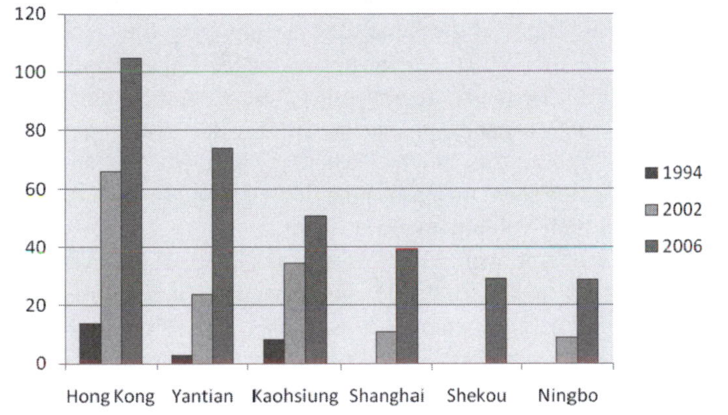

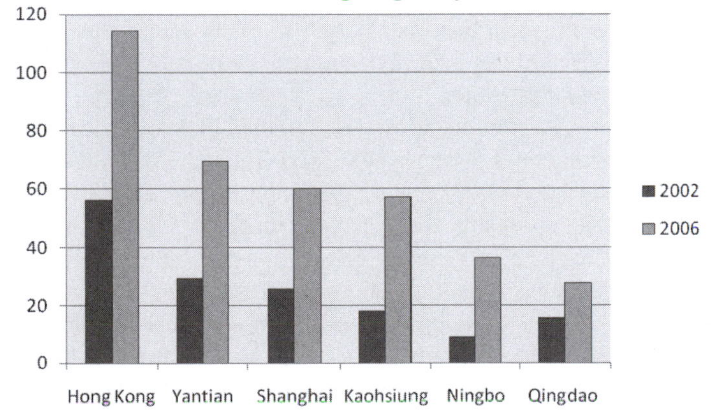

Figure 4.2 The WCTC (Thousand TEU) of Some Carriers by the Different Ports in the East Asia Range

of traffic. MSC prefers Chiwan over Yantian and Ningbo over Shanghai. Even more distinct is CMA-CGM which has a two gateway China strategy that other shipping lines only rarely use: Xiamen and Qingdao. From 1994-2002 then from 2002-2006, the network was completely redesigned to focus the development on these two gateways (Figure 4.2). On a fully growing range, CMA-CGM acted as a pioneer by going where no one is established. It is one step ahead of the competition in markets that appear to be geographically promising. In these ports, CMA-CGM has a dominating position with more than 30 per cent of the overall capacity. The same can be said about MSC in Chiwan with 37 per cent of the available capacity while Maersk has 20 per cent of Yantian's capacity.

5.3 Structuring a Network of Selected Gateways: Investing in Terminals

Table 4.4 shows major ports per shipping lines and by range. In the ports identified in the Table, a shipping line concentrates more than 20 per cent of the total capacity it possesses on that range. At the same time the shipping line accounts for more than 20 per cent of the WCTC of the port. Therefore these ports are assumed to play a key role in the maritime network of the shipping line due to the high concentration of range capacity of the line in that port and being the dominant player in the port. These ports take on "key port" status for the shipping line. For a shipping line, the dominant position is the first step to influence local port policy and also to deploy high-volume inland services.

The structuring of a network of key ports is no small affair because it demands the concentration of many efforts on one port while maintaining a dominant position. It explains why the number of key ports per shipping line varies with regards to its position in the worldwide hierarchy. Maersk with 12 ports has the largest number of key ports and has at least one key port per range. Moreover, of these 12 ports, six belong to the 25 largest container ports in the world. MSC comes second but its network is more focused on secondary ports and it does not have key ports in North East Asia. CMA-CGM is comparable. The alliances usually prefer larger ports as gateways and it is extremely difficult to achieve a dominant position in these ports.

Many differences exist from one range to another. In West Africa (Type 1), none of the sample shipping lines (Maersk, MSC and CMA-CGM) hold any major port. They tend to serve the various ports of the range in a relatively similar manner. However, they have important market shares in these ports, often surpassing 30 per cent. The West Africa range is therefore typical of a developing region: the absence of one or two main gateways and an oligopolistic situation. Since there is no real competition on the range, no shipping line needs to be the sole holder of a port as a gateway. However, Maersk tends again to act as a forerunner. In West Africa, six terminals (Apapa (Lagos), Luanda, Tema, Douala, Abidjan, Onne) belong to APMT, a subsidiary of AP Möller, the parent of Maersk. The AP Möller group reinforces its dominant position to better cope with future competitors as global terminal operators which are more and more interested in these secondary ranges.

Table 4.4 Key Ports by Shipping Line and by Maritime Range

Maritime ranges	Maersk	MSC	CMA	Evergreen	CKYH Alliance	Grand Alliance	New World Alliance
Type 1 – Introduction							
West Africa							
Type 2 – Growth							
East Asia	Yantian	Chiwan	Xiamen	Kaohsiung			
			Qingdao				
Turkey and Black Sea	Izmir	Haydarpasa					
South America East Coast	Santos	Santos					
Type 3 – Maturity							
North America West Coast	Los Angeles	Long Beach	Ensenada	Los Angeles	Oakland*	Seattle	Los Angeles*
				Tacoma	Long Beach*		
Type 4 – Advanced Maturity							
Northern Europe	Rotterdam	Antwerp					
	Bremerhaven	Bremerhaven					
North America East Coast	New York	Port Everglades			Norfolk		Miami
	Norfolk				Savannah		
	Charleston						
North East Asia	Yokohama			Osaka	Tokyo*	Kobe*	
	Gwangyang						
Southern Europe	Malaga	Valencia	Marseille Fos		Naples*		
		Barcelona	Genoa				
		La Spezia					

* Terminal investments referred to single alliance's members

Legend – Carriers' ownership in terminals

Customer (no investments) | 50/50 JV

Minority shares | Partially owned subsidiary/Wholly owned subsidiary

Note: A key port is a port in which a shipping line concentrates more than 20 per cent of its total capacity on the range and at the same time provides more than 20 per cent of the WCTC of the port. Example: on the East Asian range, Maersk concentrates more than 20 per cent of its WCTC on the port of Yantian and at the same time provides more than 20 per cent of the WCTC of this port.

Source: Authors elaboration from Drewry (2008).

On the North East Asia Range (Type 2), the consequence of the growth of the South-Korean ports and the strengthening of Busan as the major and multi-user port was the relative decline of the Japanese ports. The shipping lines that serve emerging ranges of Turkey and Black Sea and the South American East Coast are pioneers. Being one of the first lines to serve these emerging ranges helps to establish dominant position in ports and to get ahead of other competitors.

In Type 3 ranges (North America West Coast), there are few alternatives. All the attention is therefore focussed on one port complex – Los Angeles and Long Beach ports – with back and forth movements between the two. Thus Maersk left Long Beach for Los Angeles and MSC did the opposite.

In Type 4 ranges, it is more difficult for shipping lines to select a secondary port that is as strongly growing as in Type 2 ranges. In Type 4 ranges, we witness a *de facto* distribution of ports between shipping lines which doesn't exclude some crossovers. Thus, in Northern Europe, Maersk holds Rotterdam while MSC is strongly based in Antwerp. But both are also deeply involved in Bremerhaven.

In order to support such relevant maritime traffic volumes, in these key ports (Table 4.4) shipping lines often own and operate terminals, even through various forms of cooperation with stevedoring operators. Ownership patterns can range from minority shares in consortia, 50/50 equity joint ventures, to partially (POSs) and wholly owned subsidiaries (WOSs). If a shipping line is vertically integrated in port activities this does not necessarily mean that its terminals are fully dedicated to its own vessels (Olivier 2005), but generally the higher the carrier's equity involvement the more terminal resources it commands. But substantial differences among carriers' behaviours in managing intra-group transactions might emerge. This derives from the market and financial power of the company, its culture and history or the port typology. For instance, Maersk can rely on the port network of APMT even if the two subsidiaries of the AP Möller Group act as two separate profit centres and have different clients. On the contrary, the involvement of the CMA-CGM in building a terminal network is a very recent story and most of its terminals are fully dedicated. Outside Europe, the French Group has in particular decided to invest in Xiamen.

6. Conclusions

We suggest that the establishment of dedicated gateways per range is a key issue for shipping lines. But at the same time, we highlight that this goal cannot be achieved everywhere by all lines due to great differences between ranges and between shipping lines. Diversity is the rule and not uniformity. For the shipping lines, a developing market provides more port opportunities than a mature market as in the former the port hierarchy is still being established. In mature ranges, in spite of the overall stability of the distribution of traffic, concentration mechanisms have not ceased. Rather shipping line concentration is perceived in the setting up of dedicated gateways.

Differences also exist among shipping lines. Depending on their history, financial abilities and market power, they implement various strategies. The most powerful shipping lines build over the years a network of dedicated gateways. On the contrary, other shipping lines have to pursue more selective strategies because they have lesser financial capabilities or simply because they are latecomers in the container shipping industry.

The current economic downturn should change the above framework. After many years of two-digit growth, in 2008 the world throughput showed a 5.4 per cent increase only and, in 2009, we just experienced the first decline in the overall containerization history (-10 per cent). As was discussed in Chapter 3, the uncertainty surrounding economic recovery and consuming markets makes it extremely difficult to estimate the impact on liner shipping.

However, this new background will probably lead shipping lines to redesign their maritime networks. In fact, the sharp decline in cargo flows in most port ranges might prime a "re-concentration" process in the major regions. In this respect, ocean carriers are expected to adopt slightly different strategies, in accordance to the degree of involvement in terminal ownership. Latecomers in stevedoring operations should be able to easily reverse their strategies, by reducing their activities in those ports where weekly call sizes become uneconomical. On the contrary, heavily asset-based players such as Maersk and Cosco could be tempted to keep their positioning in the key areas as much as possible, hoping for a forthcoming economic recovery. Therefore, in the light of the new context, the achievement of an aggressive vertical strategy in ports might be read as a potential element of weakness because of a lower network flexibility and higher barriers of entry.

Chapter 5

Transport and Logistics Hubs: Separating Fact from Fiction

Elisabeth Gouvernal, Valérie Lavaud-Letilleul and Brian Slack

1. Shipping and Logistics Hubs: Two Functionally Different Systems Involving Different Actors

In the period leading up to the global economic crisis of the late 2000s, the development of new shipping networks and the proliferation of logistics centres were producing new forms of spatial organization. The emergence of transhipment hubs became a feature of shipping networks around the world, and the concentration of logistics activities in specific locations was a major characteristic of supply chains. Hubbing was a feature of both. The goal of this chapter is to explore how the concept of hubbing is applied and understood differently by many of those involved. We claim that the distinctions are important because there is an erroneous assumption by some that shipping hubs and logistics hubs are synonymous, and that container shipping hubs are also logistics poles. We argue that the container shipping hubs established by the carriers are functionally and frequently geographically distinct from the distribution complexes established by the logistics industry. The former is the outcome of the carriers' concerns with optimizing their route networks, the latter with managing goods distribution.

Part of the confusion arises out of the fact that both parties are concerned with optimizing flows: the logistics industry with stock and the shipping lines with containers, especially the empty boxes that have to be repositioned because of unbalanced trade flows. However, as Gouvernal and Huchet (1998) have explained, the overarching interests of the carriers are with containers themselves, which is quite distinct from the content of the containers which is made up of goods controlled by the shippers and third party agents such as forwarders. Only occasionally are the carriers concerned with the freight itself (Frémont 2007a). This dichotomy gives rise to two very different approaches to the concept of hubbing. In some cases they converge, in others they are quite distinct.

In this chapter we begin by focusing on the interests and objectives of the carriers, drawn in part by personal experience of working for a shipping line. The hypotheses presented are validated by an analysis of data obtained from *Alphaliner*, a commercial information source that monitors all maritime services. This database allows details of services of individual carriers, the ports of call used, the size and type of vessels deployed and their frequencies to be recorded. It also permits a list of all the shipping lines calling at each port to be drawn up.

We then go on to examine the hubbing associated with logistics. We demonstrate that the main actors controlling freight movements, the shippers and the intermediaries, are bound by the geography of markets in selecting the sites of logistics hubs. Evidence is drawn from a number of case studies in Europe, Asia and North America.

In the latter part of the chapter we examine the misconceptions about shipping and logistic hubbing that are prevalent, particularly in public policies that seek to promote economic development in transhipment hub ports. We argue this is ill-founded in most cases. As international trade recovers during the next decade it is essential that these misconceptions are cast aside and that more realistic assessments of the value added potential of transhipment hubs are undertaken.

2. Hubbing from the Perspective of the Shipping Line

2.1 Why Hubs?

The objective of the shipowner, or rather the carrier, since the shipping line is not always the owner of the vessels deployed on its services, is to serve a market. Its revenues are based on the transport services sold to the clients: the shippers and intermediaries. The carrier seeks above all to maximize its receipts. Gateway ports, those ports that serve a major hinterland where many customers are located, are the priority markets the carriers seek to serve. Because of scale economies in vessels, carriers strive to deploy the largest capacity ships where possible since these vessels generate more revenues while minimizing costs. The strategy therefore is to develop a service network for the major markets. This simple relationship is fundamental to the carriers' business operations. Thus, to view the objectives of the carriers as primarily to minimize the costs of operating their assets is not correct.

The ideal situation for a carrier would be to serve one port on each range to load a full vessel. This hardly ever happens, since ports cannot provide enough boxes to fully load vessels calling on a weekly service. Thus, the carriers are obliged to make several port calls on each range where and when there are sufficient containers to justify a port visit. As a result, there is a strong tendency for traffic concentration at a few hub ports. Smaller markets have to be served by smaller vessels, but in order to maximize the deployment of the largest (and highest revenue-generating) vessels, the carriers in some situations establish hubs served by mother ships to distribute the containers to the lesser regional markets. The carriers calculate the costs for serving the port directly by the mother ship and compare them with the costs of transhipment at another port and serving the port in question by feeder.

2.2 Types of Carrier Hubs

2.2.1 Gateway hubs are the principal ports of access to large market areas. They are products of the concept of centrality developed by Fleming and Hayuth (1994). They attract the carriers because of the traffic volumes available to fill ships. Examples include many of the big Asian ports such as Shanghai and Shenzhen, as well as all the major ports in North America and in Europe.

The size of the market and the possibilities of filling vessels may cause the carriers to overlook and adapt to adverse navigation or site conditions. The major European hub ports of Hamburg and Antwerp are good examples. Both suffer from being upstream ports (Baird 2002), requiring several additional hours of sailing from the ocean, equivalent to extra time in port. Furthermore, they have water depth limitations, so that access for the largest vessels has to be scheduled during the small windows of opportunity of high tide. They represent essential ports of call, however, because of the size of the markets they serve. Their hinterlands are rich and extensive. Thus, while Le Havre, another Northern Range port but with excellent nautical conditions, attracts 51 container services, Hamburg draws 92 services (http://www.axs-alphaliner.com [accessed: 27 April 2009]).

The carriers consider calls at ports such as Antwerp and Hamburg as not optional, but as essential because of their capacity to contribute large numbers of containers and maximize revenues. It is significant that, despite their geographical limitations, these two ports have enjoyed the highest growth rates in recent years among all others on the Northern Range (see Table 5.1). Examination of the service networks of the Far East-Europe carriers indicates that all the companies serve at least three ports of call on each service (see Table 5.2).

2.2.2 Transhipment hubs exploit certain geographical advantages based on intermediacy (Fleming and Hayuth 1994). These are the archetypical hubs comparable to those developed in the airline industry. They meet the hub definition of O'Kelly and Miller: " … used to denote a major switching centre in a many-to-many distribution system … the key idea is that the flow between a set of origin and destination cities (*substitute ports*) passes through one or more of the hubs, en route to the final destination" (O'Kelly and Miller 1994).

There are, however, two types of transhipment hubs. First are those whose function is to accommodate the largest vessels and tranship containers via smaller feeder vessels to other ports that either cannot accommodate the largest ships or have insufficient traffic to justify a vessel call by a mother ship (see Figure 5.1, H-2). Examples include Singapore that has achieved its status because of its intermediacy between the smaller markets of South East Asia. In the Mediterranean, the transhipment hubs are located to minimize the deviation from the shortest route across the sea between the Suez Canal and the Strait of Gibraltar (Zohil and Prijon 1999). Thus the largest container vessels in service on the Far East – Northern Range services can offload large numbers of containers at hub ports for distribution by feeders to smaller ports throughout the Mediterranean and Black

Table 5.1 Some Characteristics of Container Traffic in the Northern Range Ports (1975-2008)

	Container throughput in 1000 TEUs (2008)	% Total (1975)	% Total (1985)	% Total (1995)	% Total (2005)	% Total (2008)	% Transhipment (2006)	% Hinterland (barge, rail, road) (2006)
Le Havre	2,500	9	8	7	7	6	37	63
Dunkirk	215	1	1	1	1	1	-	-
Zebrugge	2,210	6	3	4	4	6	7	93
Antwerp	8,664	13	18	17	21	22	18	82
Rotterdam	10,784	40	39	36	29	27	27	73
Amsterdam	435	1	1	1	1	1	-	-
Bremrhaven	5,448	16	14	12	12	14	62	38
Hamburg	9,700	13	16	22	25	24	46	64

Source: Port authorities, ISL, Journal de la Marine Marchande.

Table 5.2 A Sample of Services Calling in North Europe (May 2009)

Carrier	Service	Weekly capacity (teus)	Vessel sizes (teus)	No. of ships	N. European ports of call	Hub in the Med.	Remarks
Evergreen	WCNA-Asia-Europe pendulum service	6,202	5,300-7,000	13	Rotterdam, Hamburg, Thamesport, Zeebrugge	Taranto	
Maersk	Maersk Line AE10	8,392	8,500	9	Le Havre, Felixstowe, Zeebrugge, Dunkirk, Göteborg, Aarhus, Bremerhaven, Rotterdam, Le Havre	none	Tanjung Pelapas 1st hub called at
Maersk	AE 9	6,518	6,200-7,200	8	Zeebrugge, Felixstowe, Bremerhaven, Rotterdam	Algeciras	

Table 5.2 ***Continued***

Carrier	Service	Weekly capacity (teus)	Vessel sizes (teus)	No. of ships	N. European ports of call	Hub in the Med.	Remarks
MSC	Lion service	8,240	6,700-9,200	11	Le Havre, Hamburg, Bremerhaven, Antwerp	Takes the Cape route	Antwerp (skipped on some sailings)
CMA-CGM	Asia-Europe service FAL 1	10,858	8,500-11,000	10	Southampton, Hamburg, Rotterdam, Zeebrugge, Le Havre	Malta	
CMA CGM	Asia Europe service FAL 3	8,748	8,500-9,500	9	Le Havre, Dunkirk, Zeebrugge, Hamburg Rotterdam, Zeebrugge, Southampton	none	
CSCL	Asia Europe Express 1 (AEX 1)	8,992	5,500-9,500	9	Felixstowe, Hamburg, Antwerp	none	Following ports: Barcelona, Valencia, Genoa
New World Alliance	South China Express (SCX/CEX)	7,982	6,400-8,100	9	Southampton, Zeebrugge, Hamburg, Rotterdam	none	Following ports: Port Said
Grand Alliance*	Europe-Asia – loop EU 2 (focus on S. China/SE Asia) – Via Middle East eastb/d	5,395	6,200-8,000	10	Le Havre, Antwerp, Hamburg, Rotterdam, Southampton	Cagliari	
UASC	China-ME-Europe service	5,930	3,900-6,900	10	Rotterdam, Thamesport, Hamburg, Antwerp	none	

Note: * Hapag-Lloyd/NYK/OOCL/MISC. All services are weekly.

Source: http://www.axs-alphaliner.com.

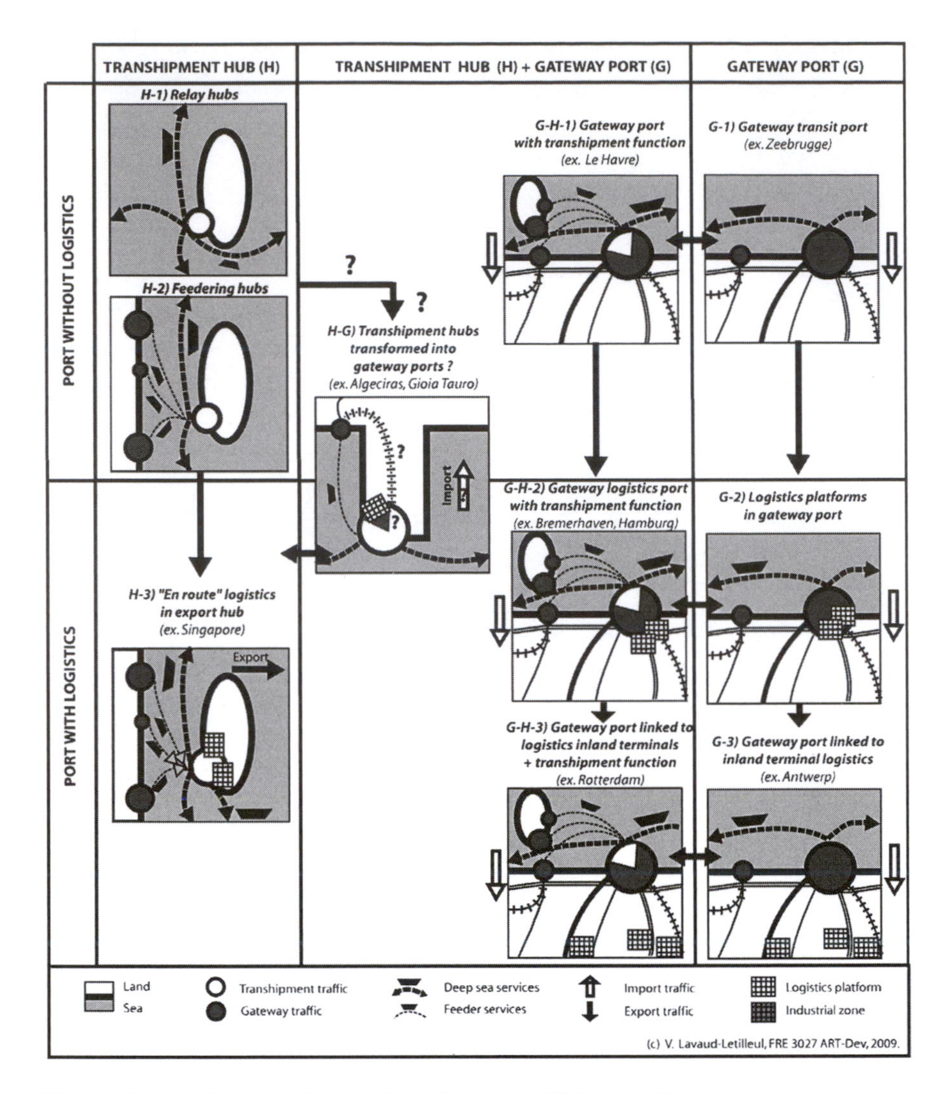

Figure 5.1 Gateway Ports, Transhipment Hubs and Logistics

Seas. Even within this category there are distinctions. Hubs such as Singapore and Gioia Tauro are open to multiple carriers, but there are others that are dedicated to a single or a restricted number of carriers. Thus, Malta, Taranto, Tanjung Pelepas, Tangiers and Cagliari act as dedicated ports.

The second type of transhipment hub is an intermediate location where containers are exchanged between mother ships on different mainline services (see Figure 5.1, H-1). These are sometimes called *relay hubs*. Maersk and MSC are major exploiters of this system. In the case of the former it has established

Algeciras as a port where its north-south services meet its east-west services (Frémont 2007a).

The transhipment hubs vary greatly in size, which is a function of the size of the ships and the volumes transhipped. Thus, while Singapore and Kingston serve the same function, one for the markets of South East Asia, the other for the Caribbean (McCalla 2008), the former handles vessels up to 14,000 TEU capacity, while the latter handles 3,000 TEU ships.

From the perspective of the carriers using ports for transhipment a major concern is the availability of adequate capacity to handle the transfers. For this reason they either invest in their own dedicated facilities, or negotiate specific agreements with the terminal operating companies for priority handling. Thus issues such as terminal productivity and terminal rates are of great importance for the carriers. A small cost difference or a short time delay in docking or cargo handling can quickly lead to an evaluation of maintaining calls at the port.

2.3 Trends in Hubbing

The taxonomy reviewed above hides a number of dynamic features. The carriers are constantly monitoring and assessing their networks and evaluating the performance of the ports they use as hubs.

Most stable are the gateway hubs. As long as there is sufficient traffic to justify a port call, carriers will include the port in their rotations. However, should market conditions change or should performance deteriorate carriers can replace a port in a particular service rotation. Several carriers have cut Felixstowe from their port rotations on the Northern Range for certain periods and have substituted feeder services from Le Havre, Antwerp and Rotterdam. Another situation is where carriers can reduce sailing time and extra time in ports by reducing the number of ports on each range, targeting in particular the ports with the lowest volumes. This is sometimes referred to as tailcutting. The Scandinavian ports, such as Göteborg, that used to be served directly on the East-West trades, are now served by feeders operating out of the port of Hamburg. In some cases former ports served by feeders can develop sufficient traffic bases to warrant direct services. This was the case of the Northern Mediterranean ports of Barcelona and Genoa that used to handle Asia traffic by feeders between Southern Mediterranean transhipment ports. The growth of traffic generated by the hinterlands of these ports resulted in direct services from Asia being established in the early 2000s. This example illustrates also the factor of vessel availability. By the turn of the century the carriers were adding newer larger vessels to their fleets. These super post panamax ships are deployed by the carriers on the main routes. This made the former panamax vessels available for deployment on new services. Thus while it remained uneconomic to use the largest vessels to call at Barcelona or Genoa, the availability of 3,000-4,000 TEU vessels coupled with the growth of markets in the Northern Mediterranean made it commercially attractive for the carriers to deploy these vessels. The carriers had calculated that direct calls for an Asia service were less costly than transhipment and feedering (Gouvernal 2002).

There is thus a trend for gateway ports to add transhipment functions. Because of their hinterland markets the carriers are drawn to them and employ their largest (and highest revenue generating) vessels. If the geographical location of a port is one of intermediacy in relation to the foreland, then it becomes commercially advantageous to organize feeder services to the smaller regional ports. Rotterdam and Hamburg as leading European gateways are today developing and extending a transhipment role (see Figure 5.1, G-H-2 and G-H-3). A further example is Valencia, a former gateway port which has been chosen by MSC to tranship containers between vessels on different services. The choices of the carriers are made not from any process that can be measured by elegant mathematical equations, rather the companies address specific constraints and seek to exploit particular opportunities as they become available.

Transhipment hubs are a different case. Because these ports have little or no traffic base of their own, their status remains highly unstable, and carriers can transfer their services to other ports with some ease. There have been shifts in the alignment of carriers in both the Mediterranean and Caribbean Seas (Gouvernal *et al.* 2005; McCalla *et al.* 2005). Once the mainline carriers remove their operations the ports have nothing to fall back on.

While there is movement of some gateway hubs to acquire transhipment functions, so far the pure transhipment hubs have not developed hinterland traffic. This is explained by the fact that they exploit intermediate geographical locations that may be attractive to the service networks of the carriers, but are not easily accessible to local and regional markets. This is particularly true of island hubs such as Malta, Cagliari and Kingston. A major question discussed below is whether this will change in the future for the hubs located on continental locations, such as Algeciras and Gioia Tauro.

3. Logistics

While the container shipping lines are preoccupied with filling ships with boxes, what is in the containers is of little direct concern to them (Frémont 2009a). The goods in the containers are of immediate interest to the shippers and customers. For them the placing of containers full of their goods on ships to be delivered from one port to another is only one part of a process of freight management that may extend from the sourcing of raw materials, their manufacture, the assembly of components and their eventual delivery to the customer. These freight management issues have become structured in what is referred to as logistics and supply chain management.

Transportation is central to logistics. Inevitably logistics providers have to incorporate various modes of transport carriers in their supply chains. In some cases they will operate their own transport systems, as exemplified by the specialized air express freight companies such as UPS, and the ownership of trucking fleets by a large number of manufacturers, retailers and forwarders. However, transport provision involving rail, sea and air modes is largely the purview of specialized transportation companies with whom the logistics firms have to negotiate rates and services. Only a few of these transport companies provide logistics services, and even of those that have a logistics subsidiary, logistics activities represent a small part of their overall transport operations (Frémont 2009a). Thus there is a major division in logistics between carriers providing transport services, and the logistics providers who are involved in organizing the cargo flows.

Logistics has grown significantly over the last 50 years, in particular the numbers of different activities that are represented (Meidutė 2005; Vogt *et al.* 2005). At the heart of logistics is inventory control and management. These activities bear relation to the traditional function of warehousing, but whereas warehousing was an intermediate activity, between supplier and consumer, inventory control today extends to manufacturing and retailing themselves, so that the entire supply chain can be ordered and managed smoothly. From these core functions have emerged other logistics activities, including order processing, invoicing, cross-docking, final assembly, packaging, labelling and customer returns. Logistics involves both the physical handling of goods flows and also the information management required to optimize the chains.

The elaboration of logistics resulted in a significant spatial readjustment. Warehousing, the basic and original logistics function, used to be characterized by the extensive distribution of storage facilities throughout economic space, with greater concentrations adjacent to major transport terminals such as ports and railway yards. Here goods of all kinds were assembled and stored, frequently in multi-storey buildings, until delivery was needed. The refinements in logistics have produced a number of major consequences (Hesse and Rodrigue 2004). First, storage was seen as a cost factor involving the unproductive use of capital, and so one of the major achievements of logistics has been the reduction of inventories and the more rapid turnaround of goods held in storage. Second, new logistics functions, such as order processing, packaging and labelling, have created added value, and thus inventory management has become an important profit centre in supply chains. Third, trade liberalization and globalization greatly increased the volumes of trade. The result of these trends is that firms have sought to reduce the number of warehouses, thereby achieving economies (less inventory, less rental costs), and centralizing inventory management and logistics at a small number of major distribution centres (DCs) that serve as regional or national markets (ESCAP 2003). For example, foreign firms serving the European market typically establish one European DC. As a result of this concentration, important logistics hubs comprising DCs of several firms have emerged. These hubs typically comprise the distribution facilities of manufacturing firms such as Sony-Ericsson, major retailers such as Wal-Mart, and third party logistics providers (3PLs) such as Kuehne and Nagel. The latter are intermediate companies that specialize in providing logistics services to clients who have opted to sub-contract out their logistics needs.

The most important locational determinant for these logistics hubs is market access (Mérenne-Shoumaker 2007; Savy 2005). In one sense this means that major centres of economic activity and population concentration possess general advantages to attract DCs, but because accessibility implies transportation, and because transportation plays an important role in logistics, ports, railway terminals and major freight airports serving major market areas represent particularly important specific location determinants for DCs. For example, more than half the European logistics centres of US and Asian companies are located in Holland, due in part to the presence of the port of Rotterdam and Schiphol airport (ESCAP 2003). The largest concentration of DCs in the world is the Inland Empire of Southern California, whose firms are tied to the ports of Los Angeles and Long Beach for container imports, as well as to the rail intermodal yards for national distribution (Bonacich and Wilson 2007). Thus good transport infrastructure AND market access are essential requirements for logistics hubs.

As Notteboom and Rodrigue (2005) and Notteboom (2008a) have noted transport terminal locations, such as ports, suffer from space limitations that result in high land costs and congestion. Modern DCs comprise very large single-storey structures, sometimes in excess of 10,000 m^2, lined with a very large number of loading bays and thus require very extensive sites. Consequently, DCs are increasingly locating on greenfield sites some distance from the port terminals, but where there is superior market access, especially with good highway connections (Mérenne-Shoumaker 2007; Leitner and Harrison 2001). These are sometimes referred to as dry ports (Roso *et al.* 2009).

These logistics hubs are drawn also to these peripheral sites because of weaker zoning restrictions in rural areas. The scale of the buildings and their lack of any aesthetic merit, as well as the traffic generated by DCs make it difficult for them to obtain planning permission in urbanized areas. The peripheral sites also make it easier to attract workers generally willing to accept the lower hourly wage rates and more flexible work schedules required in DCs. Bonacich and Wilson (2007) indicate the key role of employment agencies in these peripheral regions for the recruitment and provision of part-time workers.

The decentralization of logistics hubs from the main hinterland ports, however, is of limited geographic scale when compared with the global scope of their transportation (Van der Lugt and De Langen 2005). The DCs are still located within the major market areas, and market size continues to play a determining role in their location. The Inland Empire in California, for example, is constrained spatially by the accessibility to the ports so that truckers can make at least one pick up and delivery per day from the container terminals (Bonacich and Wilson 2007).

4. Co-Existence of Logistics Activities and Ports

In the previous sections we have demonstrated the differences between container and logistics operations. The question we explore in this section is the extent to which they can come together and co-exist in the same port complex. We have seen that gateway ports are well suited to attract logistics operations involved in opening the boxes and the storage, handling and processing of their contents. On the other hand it has been demonstrated that transhipment and logistics represent two very different practices. Nevertheless, institutional actors, including port authorities and regional development agencies among others, have frequently sought to bring the two together. In this section we will explore the reasons why public bodies seek to combine transhipment and logistics in their development strategies. We then go on to assess how realistic these goals may be.

4.1 The Interests of Public Actors in Hubbing and Logistics

4.1.1 Transhipment traffic as a means to boost container activity The universal measure of container port activity is the number of TEUs handled. Port authorities use these totals extensively in their promotional literature, and it is regarded as a measure of status. Because transhipment involves at least two lifts, this activity inflates port TEU traffic totals (Charlier and Ridolfi 1994). Thus, the actual number of different containers handled at transhipment ports is much lower than the published totals (Gouvernal *et al.* 2005). If their transhipment traffic is taken out, ports such as Algeciras appear insignificant.

Attracting transhipment traffic is an objective of many port authorities that do not have access to a large hinterland. This is particularly evident where new ports are being established, as in the Mediterranean such as Gioia Tauro and Taranto and in the Caribbean such as Caucedo and Freeport (McCalla *et al.* 2005). This also appears in established ports that have underutilized facilities, such as Dunkirk, Zeebrugge and Marseille (Frémont and Lavaud-Letilleul 2009).

For the gateway ports, transhipment traffic is an advantage but also a problem. It adds more containers to the port complex which makes it ever more attractive to the carriers. However, it contributes to terminal congestion, because these are boxes that do not leave the port, and the traffic does not contribute significantly to the value added.

4.1.2 Logistics as a means to anchor customers and create employment The container has introduced a great deal of volatility in the port industry (Slack 1993). It has permitted a high degree of traffic centralization and facilitated inland distribution by rail and waterways. This has produced significant port concentration in a few gateway ports, such as Antwerp and Hamburg. However, the growth of container traffic at a relatively small number of ports has not led necessarily to a commensurate increase in added value to the port region. Port authorities and other public actors see the development of logistics activities as a means of anchoring the region within supply chains, thus ensuring an ongoing traffic base for the carriers, and also as a means of augmenting local employment and thereby increasing the value added of container handling.

The interests of public officials must now be compared with the realities of the relationships between different types of container ports and the nature of logistics.

4.2 Relationships between Gateway Ports and Logistics

For the large gateway ports, the addition of transhipment trade can be considered as a plus because it augments the volume of containers passing through the port which enhances its attractiveness as a port of call. Thus, the mainline carriers can achieve economies of scope to the advantage of their hinterland traffic, and port authorities benefit by handling more containers and extending the ports' foreland.

However, as discussed above, public actors also wish to ensure that the port and its region benefit from the container trade, and hence it is not just a question of quantity of containers handled, but the benefits that this trade can promote. Rotterdam, the largest European container port, realized in the 1990s that its container trade was not producing the regional economic impacts of its smaller rival, Antwerp (Lavaud-Letilleul 2005). It therefore introduced a policy of establishing *distriparks* on sites within the port boundaries. This was recognition of the need to link port gateway functions with logistics activities (see Figure 5.1, G-H-3). This has been defined as the *logistics gateway* (Van Klink 1995) (see Figure 5.1, G-2). Otherwise the gateway port is simply a *port of transit* (see Figure 5.1, G-1). The problem with *distriparks* is that port sites are already limited and congested by port operations, so that the costs of land in particular, and labour act as serious constraints on the establishment of logistics activities within the port areas.

The result is that private logistics operators have been establishing their DCs away from the port complexes on extensive greenfield sites (see "Logistics" and Figure 5.1, G-3, this chapter). These sites have been established on sites with good accessibility to the main centres of population. However, the port remains a vital locational determinant, since the logistics activities are based on the receipt of maritime containers, their emptying, and the storage, packaging, cross-docking of their contents. It has led to a new form of port regionalization (Notteboom and Rodrigue 2005, 2007, 2008, 2009).

4.3 Transhipment Ports and Gateway Functions

Because of the uncertainties over the permanence of carriers calling at transhipment hubs, various actors have sought to develop gateway trade (see Figure 5.1, H-G). Algeciras is an example. For a long time this port has been the fiefdom of Maersk's transhipment activities in the region. Recently, the port authority has signed a contract with TCB, a major Spanish-owned terminal operator active in Barcelona, Cuba and Brazil. The objective is to develop a new terminal to serve the markets of southern Spain, thereby diversifying the port's activities. A rail connection is already in place. A comparable example is the port of Gioia Tauro, which since its inception as a transhipment port has sought to establish gateway traffic from the north of Italy. It has regularly requested public investments to develop this function (ISFORT 1998). More recently, in 2006, the Italian government decreed that in future the State would concentrate its investments in three Italian ports only: Genoa, Taranto and Gioia Tauro, the latter two being transhipment hubs (Debrie *et al.* 2008).

These projects present several questions. What is the likelihood of these transhipment hubs gaining gateway traffic? Algeciras, for example, has existed for more than 20 years without diversifying. The Mediterranean transhipment hubs were selected because of their maritime accessibility, but are a great land distance from major markets that are already served by national gateway ports (Valencia, Barcelona, and Genoa). What sense is there in light of present European policies favouring short sea shipping, to invest in a solution that favours land transport? Finally, if these ports remain handling containers only who would benefit from the public investments, other than the carriers?

4.4 Can Logistics be Adapted to Transhipment?

Certain public actors such as port authorities, politicians and transport providers have put forward the proposition that various logistics activities could be undertaken at a transhipment hub on goods stripped from containers prior to the cargo's onward shipment in boxes. The successful example of Singapore (ESCAP 2003), which is not only a major transhipment hub, but also one of the largest logistics centres in Asia (see Figure 5.1, H-3). Singapore developed as a logistics centre in the 1980s, with the relocation of industrial production from Japan to countries of South East Asia, where labour costs were much lower. Public authorities in Singapore (the Port of Singapore Authority and various government agencies) exploited the opportunity to tranship industrial products from Indonesia, Malaysia and Thailand to Europe and North America. In parallel, international companies were encouraged to establish their regional logistics operations in the city-state. As a result information and electronics components, among others, that were manufactured in neighbouring countries were feedered to Singapore where the parts were assembled, labelled and packaged prior to the goods being loaded into containers to be put on ships for delivery to the major consumer markets (ESCAP 2003).

The unique success of Singapore is due to a number of factors: the quality of public provided infrastructures and equipment; financial incentives; and the establishment of free trade zones which allowed for international companies to circumvent national quotas and defer payment of duties (Bruyas 2000; Wang 2009). The most important factor, however, was the high levels of education and skills of a workforce that could adapt to new technologies and work practices. This is in contrast to the skill levels of neighbouring countries (ESCAP 2003). These success factors are more or less unique to the particular case of Singapore and its geographical location with regards emerging markets, and are difficult to replicate.

Given that logistics activities overwhelmingly have favoured market locations, and given the paucity of examples of transhipment hubs attracting logistics operations, we raise finally the question as to whether it is realistic to expect logistics operations involving specific supply chains to be established at a transhipment hub. It has been suggested that a Mediterranean hub might be able to develop logistics activities (ISFORT 1998) because of a geographical location close to Europe and because of the high land and labour costs in the main EU markets. Would these cost advantages be sufficient to outweigh potential disadvantages? Certainly, the lack of diversification at the established transhipment hubs such as Malta, Gioia Tauro, Cagliari and Algeciras suggest this is unlikely. The recent plans to develop Tangiers as a transhipment port (as reported on the Tangiers port website: http://www.tmsa.ma/ [accessed: 28 January 2010]) and to establish logistics activities there may be considered at first sight to be a confirmation of this possibility.

However, there would be significant extra costs in this solution. The containers would have to be stripped and re-sorted without any scale advantages that accrue to logistics centres located in port regions serving major markets. The goods would still be placed in maritime containers, rather than cross-docked to the larger domestic containers used in North America or the road trailers in Europe. It is

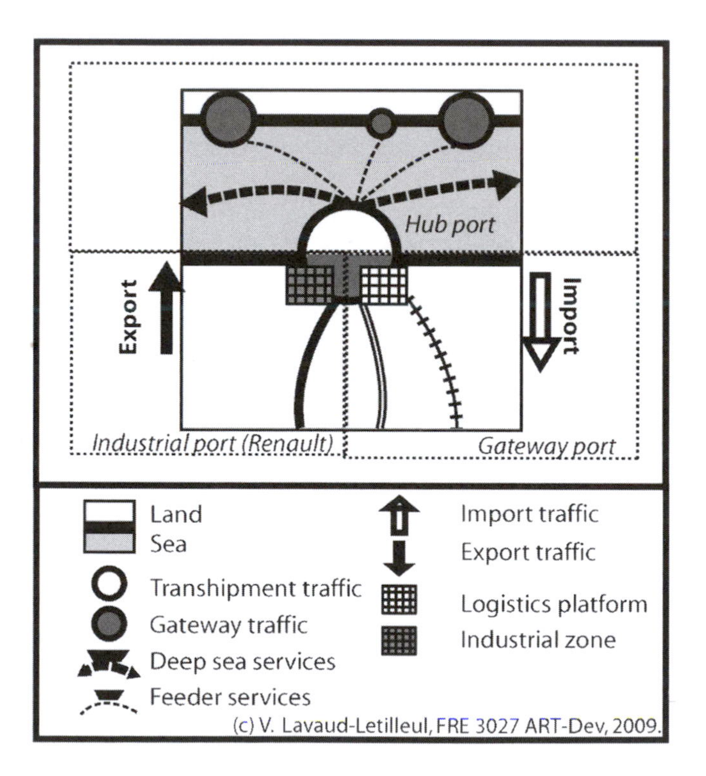

Figure 5.2 Tangiers Port Complex

probable that the assembly of final delivery batches would still have to be done closer to the markets. The Tangiers case appears to collapse when the actual plans are examined. Its organization is complex, with three main but separate functions (see Figure 5.2). The port facilities include petroleum, RORO and container terminals. The two container terminals (one dedicated to Maersk, the other to CMA CGM and MSC) have no relationship with the global distribution free zone (linked to import gateway function) and the 1,000 ha "logistics park" and free trade zone, which is more an industrial assembly/production site for Renault whose cars will be exported.

5. Conclusion

Two functionally different systems have been described, the first that of shipping lines who organize their services and manage the containers operations, and second the shippers, freight forwarders or 3PLs, that manage the logistics industry, i.e. goods distribution. The former have established transhipment hubs, the latter logistics hubs. It has been demonstrated that transhipment hubs and logistics hubs are nodes in operating chains that represent two very different practices.

Institutional actors have sought to bring the two practices together for developing port regions. This demonstrates that political factors are an important consideration in transhipment and logistics development. By considering a number of examples we have highlighted the relationships between the two approaches of the commercial actors in a political context. We suggest that while gateway ports do combine logistics functions, albeit at increasing distance from the ports themselves, it is very difficult to adapt logistics to transhipment. The exception is Singapore, but as we argue this is a very specific case and would be difficult to replicate. For Tangiers, which has been touted as another, we suggest the contrary, as there is no relationship between terminals of the shipping lines and logistics park and free trade zone. We conclude, therefore, that great caution must be employed by public agencies in seeking to promote logistics activities in transhipment ports.

This exploratory study reveals the necessity of dissecting the objectives of the different actors in transport chains. Understanding the objectives of the actors helps explain the differences between hubs. As the world economy recovers the conclusions of this study should be taken into account by academics and political decision makers alike.

Chapter 6

Port, Corridor, Gateway and Chain: Exploring the Geography of Advanced Maritime Producer Services

Peter Hall, Wouter Jacobs and Hans Koster

1. Introduction

In the most recent global economic recession, merchandise imports and exports fell even more sharply than overall economic activity. Regions that are economically highly dependent on their position within the material flows of commodities were hit hard by the decreased demand for transport. The recession revealed the strategic and economic vulnerabilities of city-regions relying upon mono-functional trade gateways. Nonetheless, despite the economic downturn, many political and policy decision-makers have kept focusing on investments in physical infrastructure in existing gateway regions (e.g. Rotterdam's Second Maasvlakte, the JadeWeser port in North Germany and Canada's Asia-Pacific Gateway Project) as a means to capture value from the ongoing forecasted growth in commodity flows. From a regional economic development perspective, what these infrastructure advocates ignore is that much of the actual value creation of these transportation flows, much of advanced service provision, and many of the decisions made about logistics investment and planning do not necessarily take place within these same gateway city-regions that accommodate the material flows. Thus, while gateway city-regions might benefit from improved infrastructure in the short run as global trade flows recover, they may simultaneously be taking a step further along a path of dependence at the expense of alternative and more valuable types of (transport-related) economic activity in the long run.

In this chapter, we explore whether advanced transportation services firms such as ship finance, insurance and maritime law, and logistics consulting concentrate within gateway city-regions, or whether they agglomerate on some other basis. The stakes in these questions are of vital importance to gateway regional economies because they concern highly skilled and value-adding activities. We argue that it will be those gateway city-regions that possess a "related variety" (Frenken *et al.* 2007) between traditional cargo-handling gateway functions and the provision of more advanced services, which, in combination with those regions' connectivity in terms of corporate decision-making provide the best prospects of economic recovery.

We construct our argument in the following manner. In the next section we review literature that has documented the spatial and economic evolution of port cities in the context of global networks. We then define and operationalize advanced services in a changing port-city context. Next we examine the North American case using quantitative data on the location and network connectivity of specialized maritime advanced service providers. We also present evidence of the spatial division of labor in the US port-logistics transportation sub-sector. While our quantitative data are rough, they do allow us to address future avenues for research and strategic policy-making. We then present a grounded case study of Vancouver, Canada, in which we explore the addressed relationships in more qualitative terms. We end with some tentative conclusions and implications for strategic policy.

2. Port Cities Within Global Flows and Networks

The organization of the shipping industry, the services which they employ, and the logistics chains they serve shape the spatial distribution of the value created by seaport gateways and their associated transportation systems. Physical separation of seaports from urban centers is a well-known and much-studied phenomenon (Hoyle 1989; Levinson 2006; Ducruet and Lee 2006; Hall 2007a). From a functional-economic point of view, city and port have become less dependent on each other as containerization reduced local demand for labor while a regionalization process pushed distribution and logistics activities into the hinterland (Notteboom and Rodrigue 2005). Likewise, Slack (1989b) advanced the hypothesis that the maritime service industry, traditionally small, fragmented and churning but typically identified with a specific close-to-the-port industrial district, faced the threat of being displaced by waterfront and downtown redevelopment. But what about wider potential economic connections between (maritime) transport activity and related, though more specialized, *advanced service provision* such as finance, insurance and law?

Previous research has suggested that maritime services are more likely to follow the locational patterns of other advanced services (see Slack 1989a; O'Connor 1989). For example, while most cargo-handling port activity in the Bay Area moved from San Francisco to Oakland in the 1970s, most advanced maritime services remained in "The City" (Campbell 1993). Similarly, while the Port of London missed out on containerization, the City of London remains the world's leading center in advanced maritime services, largely due to the spatial proximity of higher order banking, insurance and legal services (Jacobs *et al.* 2010). This observation seems to confirm the spatial-economic model of Fujita and Mori (1996), which predicts the continuing prosperity of port cities even after their initial advantage of deepwater-access becomes irrelevant. Such places continue to prosper if they manage to diversify their economic structures and evolve into agglomerations of service-orientated business and corporate decision-making. We think that these ideas and related questions about the relationship between ports and services deserve re-examination now for three related reasons.

First, there is the issue of the role of ports in global supply chains (Robinson, 2002; Wang *et al.* 2007; Jacobs and Hall 2007). Recent studies focused on *the port as a physical manifestation* of the logistical functions that these locations serve in the overall *global trade in commodities*. This literature applies conceptual insights from economic geography and supply chain management to highlight the role of port clusters and the transport sector in the "new international division of labor" that emerged during the 1980s. While these port studies and sector-specific analyses of commodity chains are yielding useful insights, they have tended to neglect the *organizational structure* behind these activities as well as their spatial connotations. Hesse (2010) has argued that the relationships between material flows, gateway infrastructure and their organizational space should be understood as an outcome of the continuous friction between "site" and "situation". Here, site refers to "local underlying area conditions" (for instance, fertile land, infrastructure or availability of skilled labor) and situation refers to the *connection* between areas and the impact they have on each other. Advanced maritime services may be regarded as site-shaping elements that provide needed resources, but they are also situation-shaping elements that provide corporate, informational and financial connections. Hence, in this view, it is increasingly important to look beyond the goods handling sector, narrowly defined, to understand the likely development trajectory of the gateway city-region.

Second, related to the rise of more complex supply chains are processes of port rescaling and regional urbanization. While ports have been separated from the cities that hosted them, in developed countries most ports are still metropolitan activity nodes. So while ports have separated from the city in physical space, they remain part of urban commute-sheds, and have become implicated in region-wide local distribution systems that connect the waterfront to suburban distribution centers, warehouses and railheads that funnel goods from ports into trade corridors. We need to consider the possibility that maritime services may co-occur with other port-logistics activities in the same extended metropolitan gateway regions.

This leads us to a third, policy-driven reason for this re-examination. Trade gateways are nodes of intense transport activity linking continental corridors and maritime (and air) trade routes. This definition of the gateway focuses on its role in physical goods movement, and it has been well established that cargo movements are concentrated in a relatively small number of gateways. Policy makers and local communities care deeply about whether this concentration of cargo-related transport activity is associated with more high-value advanced service activity. Do maritime advanced producer services cluster in gateways? If maritime advanced producer services do not cluster there, it may be harder to win local support for trade expansion.

We cannot answer these questions definitively within the scope of this chapter; rather, in what follows we aim to explore this issue at various spatial scales, using a variety of data sources.

3. Defining Advanced Services in a Changing Port-City Context

Traditional conceptions of port and port-related services have focused on services to ships (classified with Water Transportation under the old Standard Industrial Classification (SIC) system) and services to cargo (classified with Services to Water Transportation Industries under the SIC system). Under the new North American Industrial Classification System (NAICS), the cargo service activities are combined in the new category, "Support Activities for Transportation". The newly redefined sector includes most of what has traditionally been included in the port service industry (Slack 1989b; O'Connor 1989), including port and harbor operations, marine cargo handling, navigational services to shipping, other support activities for water transportation (chandlers, surveyors, etc.), and freight transportation arrangement.

The NAICS system represents an improvement over the SIC system because it combines freight- and cargo-related services for all transportation activity in a single category. This change is an important recognition that freight activity is increasingly integrated across modes. Despite this improvement, the NAICS system, like the SIC system, still gives more prominence to freight carriage than it gives to the value chains in which the freight and its carriers are embedded. This could lead to the (incorrect) presumption that freight handling and service activities are necessarily closely related. As we will argue, this is the case only for the most routine maritime services, and the relationship is not linear. Rather, we agree with O'Connor (1989) who observed that the most valued port service activities which require non-routine and high-level face-to-face interaction are better understood as a sub-category of advanced producer services.

Within the SIC system as well as within the NAICS system, specialized maritime services in the finance, insurance and legal sectors are not identified separately from other advanced services. We still have further to go to truly appreciate the implications of supply chains for ports. Conceptually, we combine insights from two related theories to clarify our understanding of the advanced maritime services sector. First, the global production network (GPN) approach pays attention to both physical supply chains and financial, contractual, informational and organizational value chains. The GPN approach asserts that these physical and value chains are combined in non-linear, networked systems of business organization, and hence they are not necessarily in geographic proximity. They are of consequence to regional development because value is created and added within them (Coe *et al.* 2008; Coe *et al.* 2004).

A second concept we employ is Massey's (1979) notion of the spatial division of labor. Massey suggested that firms seek to locate routine functions in sites of lowest cost production, primarily expressed in terms of labor costs. However, non-routine command and control, and research and development functions locate where they can access specialized skills and services and in close proximity to other decision-makers. These insights have contributed to a vast literature on firm location; the core insight is that firms may choose to divide their activities spatially on the basis of production (labor) cost. It is not clear what is the most routine part of the logistics chain; arguably it is not the cargo handling/port function. Rather, it may be in the back-office administrative activities that support transportation activity. It is hence worthwhile to explore the location of port-logistics activity from a spatial division of labor perspective.

The forgoing discussion suggests that the various elements of the advanced maritime producer services sector may not necessarily co-locate in gateway city-regions. The highest order service functions, including Protection and Insurance (P&I) head offices, decision-making and maritime specialist expertise functions of fixed premium insurers and maritime finance, are expected to locate in the key global financial centers of London, New York, Tokyo and, more recently, Singapore. In contrast, we do expect routine services to locate in port cities. In addition to the traditional port services sector (chandlers, brokers, forwarders, etc.), examples of routine, port-adjacent activities in advanced maritime services include local representatives of global cargo and vessel surveyors and local insurance correspondents.

Our deductive reasoning also suggests that there may be an intermediate category of advanced maritime service activity that is more decentralized but is not necessarily tied to the port in the way that most routine services are. These include logistics consulting, engineering, research and development functions, and ICT. We also place maritime legal in this category. Although BIMCO (the Baltic and International Maritime Council) regulations give a jurisdictional advantage to London and New York, correspondents who are also expert in local legal jurisdictions are required in every port and especially in the major gateway ports. They combine local knowledge with global connections to provide a regular, but variable and specialized service.

4. The Big Picture in North America: Measuring Advanced Maritime Producer Services

In this section we assess whether the available evidence supports the hypothesized patterns of advanced maritime services location. We make use of two different databases to explore issues of hierarchy and clustering in the advanced maritime services industry. The first is the World Shipping Register, which includes the location of maritime advanced service providers (insurance, law, insurance brokerage, surveying, consultancy) on a global scale. Although we acknowledge that this database is not complete, it does provide a valid representation of the spatial distribution of some segments of advanced maritime producer services (hereafter, AMPS). These data only overlap slightly with our second database, a representative annual survey of workers conducted by the US Census Bureau. Since these data are classified according to the SIC system, marine surveying is the only industry sector included in both analyses. We explored these data for evidence of a spatial division of labor-type hierarchy in the port-logistics sector.

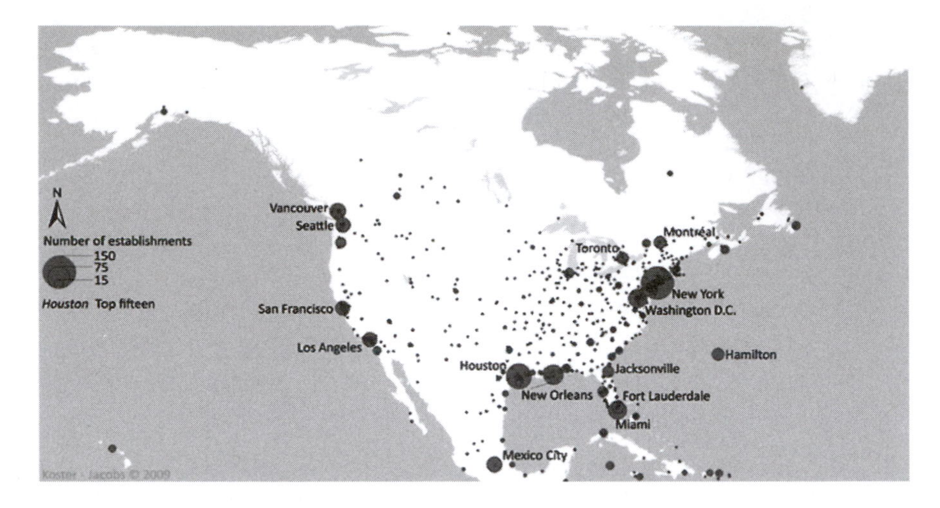

Figure 6.1 The Location of Maritime Advanced Producer Services North America

4.1 Location and Network Connectivity of AMPS

In Figure 6.1 we map the locational pattern of AMPS firms in North America. New York is ranked highest in terms of establishments followed by Houston. The relatively weak position of cities on the US West Coast is noteworthy. Although ports on the US West Coast (Los Angeles, Long Beach, Oakland, Seattle and Tacoma) and Vancouver in Canada, rank among the busiest in the world in terms of container throughput, apparently they do not host many AMPS. Instead, AMPS tend to concentrate on the US East Coast (most notably New York-New Jersey, Philadelphia and Washington, DC) and on the Gulf Coast (most notably Houston, New Orleans and Florida). This geography reflects the sheer mass of financial and human capital in New York, the location of the federal government in Washington, DC and the huge size of pleasure boating in Florida. Both Houston and New Orleans host extensive bulk-related ports. In Houston, for example, we expect strong connections (i.e. related variety) with the energy and off-shore oil sectors, especially in terms of consulting engineering and ship inspection. One large Norway-based ship inspection and consultancy firm confirmed that their Houston establishment acts as the American head office for energy related activity.

Note that some locations host significant numbers of AMPS firms, but do not host significant seaport activity (for example Ft. Lauderdale, Hamilton, Washington, DC, Toronto, Mexico City). Hamilton (Bermuda) as part of the British Commonwealth is home to some London-based P&I Clubs, which moved their offices after the devaluation of the pound sterling in the early 1970s. However, despite the number of establishments in Bermuda, these tend not to be large employers. The CEO of one London-based P&I Club domiciled in

Bermuda indicated that of the company's total 170 employees, only four work at the Bermuda office which is responsible for underwriting renewals, settling large claims and issuing contracts. All other activities, such as reinsurance, accounts and regular claims, are handled in London. The strong positions of Washington DC and Mexico City may be related to the proximity to government functions such as maritime administration. Also, AMPS seek proximity to agglomerations of general advanced service provisions (finance, insurance, law) found at the top of national urban hierarchies (Toronto, Mexico City, New York).

Although mapping advanced services provides a first perspective on its geography, it does not tell us how these cities are connected with each other through intra-firm business networks. The external network relationships of firms operating in local clusters influence their competitiveness by providing them access to flows of capital, information and new technologies generated elsewhere (see Coe *et al.* 2004; Wolfe and Gertler 2004; Markusen 1996). As Beaverstock *et al.* (2002: 114) put it, "global capacity is not a function of the size of a city's business community, but of its connectivity". In addition, looking at the intra-firm business networks between cities will allow us to identify possible patterns of hierarchy based upon corporate divisions of labor (in terms of routine based activity vs. command and control). Thus, while a location might host many AMPS establishments, these offices might conduct purely routine-based activities that serve only local markets. In order to identify the connectivity of these centers of AMPS, we adopt the network methodology as advanced by scholars in the field of World City Network research (Taylor 2001; Taylor *et al.* 2002; Derudder and Taylor 2005).

We use firm information to measure the network between cities that host AMPS firms based upon office locations. The starting point of the network analysis is a matrix with the so-called service values (v_{ij}) of firm j in city i. The bigger or more important the establishment is, the higher the service value will be. Different methodologies are used when determining the service values. We adopt the methodology of Taylor *et al.* (2002), with some adjustments. When there is no office present in city a of firm j, zero is assigned. To subsidiaries without any special command functions a score of 1 is given. Headquarters of firms with 15 or more establishments is scored 5. Regional headquarters of such firms score 3. Headquarters of a firm between eight and 14 establishments are scored 4, between four and seven establishments score 3, and headquarters of a firm with two or three establishments score 2. Taylor *et al.* (2002) point out that such data creation is far from perfect because the information collected from websites and in the annual reports often do not demonstrate a clear picture of the hierarchical structure of a firm, but that these problems will not lead to too much uncertainty since the scoring method is very simple and the data typically includes a large number of firms, reducing the impact of any errors.

Once we know the service value of each city, we can estimate the strength of the relationship between them. Hence the relationship, or elemental interlock link ($r_{ab,j}$) between two cities, a and b for firm j is:

$$r_{ab,j} = v_{aj} \cdot v_{bj}$$

The aggregate city interlock link are the summed relations between a and b of all firms located in city a and b. This is defined as:

$$r_{ab} = \sum_j r_{ab,j}$$

We divide this city interlock link by the largest city interlock (in our case New York-Washington) to arrive at the relative city interlock link. From this we can determine the situational status of a city within the network. For every firm present in a city we multiply the service value with the service values of all establishments in all other cities. Derudder and Taylor (2005) point out that this is a proxy for the city's global network connectivity (GNC):

$$GNC_a = \sum_i r_{ai} \quad a \neq i$$

Finally, to make this measure more easily interpreted, we divide each GNC by the maximum GNC, in our case Houston.

In terms of network connectivity, Houston and New York top the list followed at some considerable separation by Atlanta (see Table 6.1). While Atlanta hosts relatively few establishments and only some headquarters, Atlanta-based firms are connected with relatively many establishments in other cities. The strong GNC of Chicago can be explained by the presence of many headquarters. This does

Table 6.1 The Network of Maritime Advanced Producer Service in North America

Rank	City	Country	Relative Connectivity
1	Houston	USA	1.0000
2	New York	USA	0.9544
3	Atlanta	USA	0.6227
4	Mexico City	Mexico	0.5175
5	San Francisco	USA	0.4379
6	Miami	USA	0.4086
7	Chicago	USA	0.3987
8	Vancouver	Canada	0.3880
9	Washington, DC	USA	0.3663
10	Toronto	Canada	0.3467
11	New Orleans	USA	0.3466
12	Los Angeles	USA	0.3134
13	Montreal	Canada	0.2746
14	Portland, OR	USA	0.2630
15	San Diego	USA	0.2233

not necessarily imply that the Chicago office is directly involved in maritime services (we cannot verify this with the available data), but it does imply access to specialized advanced services for Chicago-based shippers and carriers. In other respects, the hierarchy in Table 6.1 closely matches that revealed in Figure 6.1. Seattle, Fort Lauderdale, Jacksonville and Hamilton are displaced by Atlanta,

Table 6.2 Most Important Urban Links in North America in Terms of Intra-firm Networks

Rank	Link	Relative Linkage Strength
1	New York – Washington, DC	1.0000
2	Chicago – New York	0.7143
3	Chicago – Washington, DC	0.6633
4	Miami – New York	0.5918
5	Houston – New York	0.4184
6	Miami – Washington, DC	0.4082
7	Chicago – Los Angeles	0.3980
8	New York – Philadelphia	0.3980
9	Houston – New Orleans	0.3878
10	Los Angeles – New York	0.3878
11	Mexico City – New York	0.3673
12	Houston – Mexico City	0.3571
13	Chicago – Houston	0.3469
14	Montreal – Toronto	0.3469
15	New York – San Francisco	0.3469

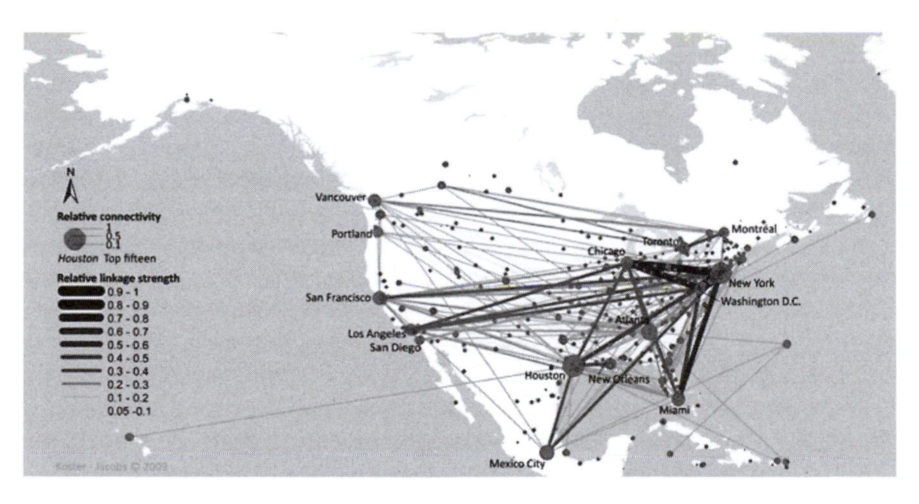

Figure 6.2 The Network of Maritime Advanced Producer Service in North America

Chicago, Portland and San Diego, indicating that AMPS firms in the former locations are relatively small and not well connected (or in the case of Hamilton, these connections are predominantly with London).

Data on inter-urban network pairings in North America in terms of intra-firm AMPS reveal the dominance of New York: 8 of the top 15 inter-urban networks include New York (Figure 6.2 and Table 6.2). Chicago also features prominently in the strongest pairings because it hosts headquarters of firms that maintain offices with high service values in New York, Washington, DC, Houston and Los Angeles.

To analyze the division of labor between more routine-based and command-control locations we again turn to a slightly modified methodology of Taylor (2001). When the service value of a firm is 1, we *assume* that this office is a routine-office, while when the service value is higher than 1, we assume that this office has some coordination and command functions. For example, in city X we can find a headquarters with a service value of 5 and a subsidiary with some coordination functions with a service value of 2. Then the "command-value" for city X is 7. When we apply this reasoning to our data, a somewhat different ranking appear (Table 6.3 and 6.4).

New York and Houston again top the list in terms of command-value. However, when we compare Table 6.3 with Figure 6.1 and Table 6.1 some differences become clear. Cities such as Chicago, Philadelphia, Toronto, and Montreal appear to rank higher in terms of command values than in total establishments and than in global network connectivity. When we compare these command values with the ranking of routine establishments (those with service value 1), then we see that New York and Houston again top the list (see Table 6.4).

These insights, although tentative, seem to indicate that New York and Houston act as the leading North American centers in terms of AMPS. Both cities dominate in terms of total establishments, intra-firm network connectivity, and command values and routine offices. The strong presence of routine offices might be based

Table 6.3 Top Ten Centers of Command Offices

Rank	City	Country	Command-Value
1	New York	USA	75
2	Houston	USA	47
3	Washington, DC	USA	33
4	Chicago	USA	32
5	Philadelphia	USA	23
6	Toronto	Canada	20
7	Miami	USA	18
8	Montreal	Canada	17
9	Los Angeles	USA	16
10	New Orleans	USA	15

Table 6.4 Top Ten Centers of Routine Offices

Rank	City	Country	Routine Establishments
1	New York	USA	135
2	Houston	USA	83
3	New Orleans	USA	54
4	Washington, DC	USA	50
5	Miami	USA	50
6	Vancouver	Canada	39
7	Seattle	USA	37
8	Mexico City	Mexico	36
9	Los Angeles	USA	28
10	San Francisco	USA	28

upon their seaport and cargo handling activity, whereas their lead in command activity seems to be based upon urbanization externalities such as proximity to general advanced services and government functions. If we compare Table 6.3 and 6.4 further, we observe the stronger presence of West Coast centers such as Seattle, Vancouver and San Francisco in routine service functions. What also becomes apparent is the entire absence in advanced services of such major US East Coast (container) ports as Hampton Roads, Norfolk, Savannah and Charleston. These places lack a related variety of economic functions and are heavily dependent on conventional cargo handling activity.

4.2 A Spatial Division of Labor in Port-logistics Industry?

In this section we explore whether there may be a spatial division of labor in the port-logistics industry, with a specific goal of trying to identify whether command-and-control occupations and high-level skills are concentrated in particular types of gateway city-regions. While this analysis builds on the foregoing section, we need to acknowledge up front that the following analysis is limited because the AMPS firms identified above are, for the most part, not included in any direct transport-providing sub-sector. This is despite the redefinition of "support service for transportation" in the NAICS system. The data for this section of our analysis includes only the United States.

Spatial division of labor and GPN theory suggests that global service cities should have more management occupations, while routine manual/handling occupations would go to cargo gateway city-regions. To explore these notions we first identify classes of Gateway Cities according to their AMPS and cargo character. For ease of communication we refer to these as "cities" although our data is based on metropolitan statistical areas, thus reflecting "city-regions". We then identify port-logistics sector workers; and finally, we compare the occupational profile of the port-logistics sector in different Gateway city-region classes.

Gateway city-regions were identified out of a combination of cargo handling and AMPS data (see Table 6.5). Note that we focus only on the very top of each hierarchy – just the top 10 container ports, top eight bulk ports and top 11 AMPS cities:

- Global Gateway Cities Tier 1 – New York and Houston stand out as major cargo port cities that are also far ahead of all other cities in terms of AMPS.
- Global Gateway Cities Tier 2 – although they are all major port cities with significant AMPS, they do not have the same AMPS activity as the Tier 1: Los Angeles-Long Beach, Seattle-Tacoma, San Francisco-Oakland, New Orleans-Baton Rouge.
- Cargo Gateway Cities – these are major port cities but they do not figure on the AMPS list: Savannah, Hampton Roads, Charleston, Corpus Christie, Beaumont.
- Service Gateway Cities – these are cities which never had, or no longer have significant cargo ports: Washington-Baltimore, Miami-Fort Lauderdale, Philadelphia, Chicago, Atlanta (noting that the ports in Baltimore, Miami and Philadelphia do maintain some cargo handling activity alongside significant commercial waterfront redevelopments)".
- All other places.

Table 6.5 Identification of Global, Cargo and Service Gateways

City Class	Port	Top container port *	Top cargo volume port **	Top AMPS city ***
Global Gateway Tier 1	New York and New Jersey	Y	Y	Y
	Houston, Texas City	Y	Y	Y
Global Gateway Tier 2	South Louisiana, New Orleans		Y	Y
	Oakland-San Francisco	Y		Y
	Seattle, Tacoma	Y		Y
	Los Angeles, Long Beach	Y	Y	Y
Cargo Gateway	Savannah	Y		
	Hampton Roads	Y		
	Charleston	Y		
	Corpus Christi		Y	
	Beaumont		Y	
Service Gateway	Washington-Baltimore			Y
	Miami-Fort Lauderdale			Y
	Philadelphia			Y
	Chicago			Y
	Atlanta			Y

Note: * In 2007, 1.5m TEU +, including foreign/domestic, loaded/empty. ** In 2007, foreign trade 50MT +. *** Top AMPS in North America by establishments and/or network connectivity.

Our employment data source is the March Annual Demographic Survey of the Current Population Survey, used previously (Hall 2009) to study relative earnings in major port cities. Due to changes in the industry classification system, we only used the 2003-2006 data. The port-logistics sector consists of the Water Transportation, Truck Transportation, Warehousing and Storage, and Support Activities for Transportation sub-sectors. The last of these includes most of what has traditionally been included in the definition of the port service industry, for example, stevedoring, brokering, and forwarding. Since the size of the trucking sector dominates the other sub-sectors we report the analysis with it both included and excluded.

Table 6.6 compares the occupational profile of port-logistics workers in four Gateway City classes, and all other places. Occupations were defined in three broad categories: (1) CONTROL includes management, engineering and advanced service functions; (2) SERVICE includes office and clerical, as well as routine service functions (cleaning, security, etc.); and (3) MANUAL includes all craft, repair, manual and transport operation and handling occupations. The following tentative conclusions emerge.

First, while all four classes of gateway city-regions have a higher proportion of control occupations in transportation than all other cities, it is hard to see a pattern beyond this. Surprisingly, Global Gateway Tier 2 and Service Gateways have slightly higher proportions of control occupations than Global Gateway

Table 6.6 Occupational Profile of US Port-logistics Workers by City Class, Without and With Truckers, 2003-2006

			5 Gateway city classes			
	Occupational group	Other	Global Gateway Tier 1	Global Gateway Tier 2	Cargo Gateway	Service Gateway
Water, Warehouse and Transportation Support	CONTROL	14.3%	20.9%	23.5%	18.4%	23.1%
	SERVICE	27.8%	33.5%	26.6%	49.0%	28.8%
	MANUAL	58.0%	45.6%	49.9%	32.5%	48.1%
Total		100.0%	100.0%	100.0%	100.0%	100.0%
Water, Warehouse, Transportation Support and Truckers	CONTROL	7.9%	11.4%	11.9%	9.5%	11.6%
	SERVICE	16.4%	22.0%	18.6%	26.7%	21.5%
	MANUAL	75.7%	66.6%	69.5%	63.8%	66.9%
Total		100.0%	100.0%	100.0%	100.0%	100.0%

Note: CONTROL includes management, engineering and advanced service functions; SERVICE includes office and clerical, as well as routine service functions (cleaning, security, etc); and MANUAL includes all craft, repair, manual and transport operation and handling occupations.

Source: Author's analysis of Current Population Survey (March Annual Demographic Files), 2003-2006.

Tier 1 cities. At the very least, this suggests that New York and Houston do not have some special advantage within the port-logistics sector itself, unlike what was found when focusing on the AMPS data. Second, service occupations form a surprisingly high proportion of the port-logistics jobs in cargo gateway city-regions; this leads us to speculate that these routine jobs are perhaps the low level occupations in port-logistics.

It is important to emphasize again that these data prevented us looking at AMPS directly; rather, they allowed us to ask whether the port-logistics sector itself displayed a spatial division of labor that followed the AMPS or cargo pattern more closely. The findings indicate that Global Gateway and Service Gateway cities overall did have more command-and-control workers than all other cities. In contrast, Cargo Gateway Cities, those with major ports but little AMPS, did not reach the same high command and control level. These findings – that Cargo Gateway cities do not sit decisively at the top of a spatial division of labor – are consistent with Hall's (2009) finding that major ports in the US do not confer a major labour market advantage on transport workers in their host cities. This also implies that the economic spin-offs of major investments in gateway infrastructure may not be captured in the gateway city-region itself.

5. The Case of Vancouver

Although not included in our US employment data, Vancouver (Canada) is an appropriate site to address qualitatively relationships suggested quantitatively above. First, Vancouver acts as gateway region for North America – Pacific Asian trade. Its terminals handled about 2.5 million TEUs and 115 million tons in 2008. Although port operations in Vancouver also became physically separated from the city center through the development of the Deltaport, intensive cargo handling activity still takes place at the docks nearby downtown at the Burrard Inlet (such as the DP World container terminal) and along the Fraser River. The historical port-city interface can still be observed in downtown buildings such as the Marine Building, completed in 1930. Secondly, Vancouver and its ports act as a crucial transport node in the national infrastructure upgrading policy, the *Asia Pacific Gateway and Corridor Initiative* (Canada 2006) that aims to improve Canada's competitive position in North America-Pacific Asia trade through a series of road/rail separations and highway and intersection improvements. While these investments in hardware might attract more cargo to Vancouver ports, it remains important to ask to what degree these efforts will imply more high-valued added activity for the urban region. Thirdly, the city and private stakeholders engaged in a pro-active cluster strategy during the 1990s aimed to attract head offices of international shipping companies and advanced (maritime) services under the banner *International Maritime Centre* (Weibe 1993). Although Vancouver ranks among the top 10 in North America in AMPS establishments (Table 6.1), it is relatively weak in terms of network connectivity and hosts mainly routine based offices (Table 6.1 and 6.4).

These policy actions and investments in both transport infrastructure and advanced services can, in combination with our quantitative data analysis, provide more insights about the spatial-economic inter-relationships between port and city. In-depth interviews about firm location dynamics were conducted with senior representatives of Vancouver-based shipping and AMPS firms. Although these interviews are only suggestive of trends, we do draw the following observations:

- Seaport activity does generate demand for more routine-based advanced services in the form of ship inspection, classification, surveying and maritime law. These businesses keep an office "in town" to be close to port whenever ships call after their international voyages (for inspection or classification) and/or when there is damage to ship, cargo or port infrastructure (maritime law, surveyors). Although these are highly valued and may involve high education levels (e.g. polytechnic or university level engineering), they tend to be routine and non-moveable. As such, these activities seem to be most *sticky* to seaport location since they provide services in direct and real-time demand.
- The attraction of global leading shipping company head office(s) does not quickly translate into significant growth in local AMPS. While fiscal incentives may have influenced the relocation of the head office of firms such as Teekay Corporation (which specializes in oil-related shipments) to Vancouver, it is also clear that proximity of the seaport and proximity to competitors or the supply of expertise (both technical and managerial) in shipping or related advanced service were not important factors. So while head office attraction does generate direct, high-paying extra jobs to the local economy, the *territorial embeddedness of such firms* in the "gateway" itself may be more limited.
- There is a positive relationship between head offices of ship owning companies and the upgrading of maritime advanced services offices. For example, a global ship classification, inspection and consultancy firm confirmed that the move of Teekay Corporation headquarters to Vancouver did imply an upgrading of their own office in Vancouver. Before Teekay's move, they acted as a liaison office to service the local shipping market generating six to eight employees. After Teekay's move, the office upgraded to 12-15 employees acting as the "Pacific Coast district management office". However, we must realize that its dynamics are highly contingent on the specific historical relationships between the firms involved.
- Firms providing AMPS had a mixed reaction to the prospect of the Asia-Pacific Gateway project attracting more cargo. In terms of routine-based activity in the Vancouver metro-region, increased traffic generated by infrastructure upgrading implies more opportunities for routine activities such as inspections, classifications and maritime arbitrage and correspondence with

> insurance companies. On the other hand, the gateway function certainly will not automatically generate increased demand for more command-value based advanced service provisions. Instead, these activities will still be undertaken at a distance either in Houston, New York, London or Singapore.

Finally, we note that traditional freight service activities in Vancouver are now widely dispersed across the metropolitan area in much the same way as are freight handling activities. In a 2007 commercial database of port-logistics establishments (it should be noted that this database did not include legal and finance firms), we identified 772 maritime and air transportation, terminal operating, trucking, warehouse and storage, freight logistics (including forwarders, brokers, and consultants) establishments. Figure 6.3 shows the location of these establishments in the Lower Vancouver Mainland. While the 252 freight logistics establishments are typically smaller than the other port-logistics establishments, they were only slightly more likely to be located in the traditional core port-city interface of the Burrard Inlet area (17.5 per cent of freight logistics establishments were located here, as opposed to 15.3 per cent of all port-logistics establishments).

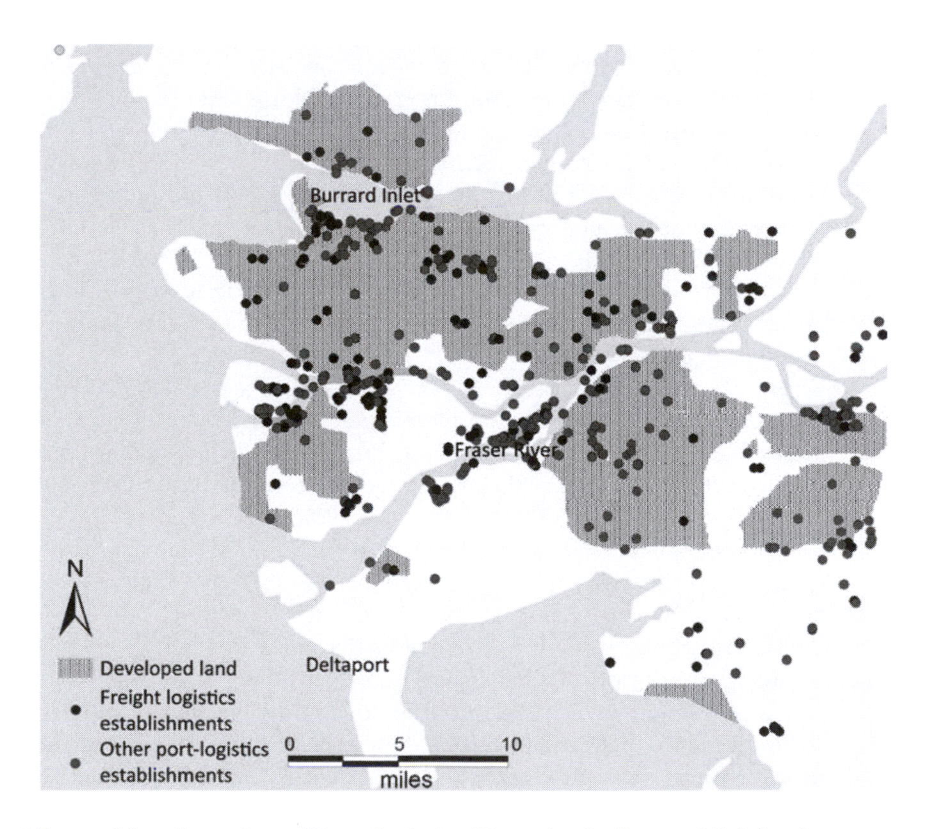

Figure 6.3 Location of Port-logistics Firms in the Lower Mainland

Our case study leads us to the tentative conclusion that Vancouver lacks a vibrant geographic cluster of advanced maritime services, and that traditional (and mostly routine) port service functions are dispersed across the metropolitan area in much the way that cargo handling activities are. This finding supports the notion that there is no necessary proximity between AMPS and freight gateways. This is a challenging finding for policy-makers; the agglomeration of advanced services and control-based transport sector activity is a local self-reinforcing, but highly contingent, process.

6. Conclusions and Policy Implications

This chapter explored the geography of advanced maritime services in North America, and asked to what extent these high valued, transport-related services agglomerate and cluster in gateways. We do not provide definitive answers to this question; rather we have explored this relationship at a variety of spatial scales, making use of both quantitative datasets and a grounded case study. Nonetheless, we are able to draw some tentative conclusions from which we can draw some policy implications.

Maritime services encompass a diversity of activities that range from the routine to the highly advanced and from the dispersed to the highly concentrated. We have tried to impose some conceptual and analytical order on the different service activities that are associated with port and transport related sectors, recognizing that the sector is constantly evolving in relation to the global reorganization of both the maritime shipping industry and global production networks.

The spatial relationship between maritime advanced services and physical gateways is limited. It is possible to have places with a concentration of advanced maritime services that have no or little port activity; in these places, the presence of advanced maritime services is more generally associated with the historically evolved agglomeration of finance, insurance, legal activities and with the proximity to maritime institutes, government functions and international branch associations. It is also possible to have major port gateways that display very little evidence of higher order, command-and-control service functions in the maritime industry. Gateway city-regions do have port service employment, but there is no need for this to extend beyond relatively routine functions, leaving some gateway economies highly dependent on physical flows. There are a small number of privileged gateways that act both as major physical nodes in global commodity flows and host many command and control functions of advanced services. While such co-location might suggest a process of co-evolution, it is more likely that this is the coincidence of two unrelated dynamics. However, future research should test these variables in a more rigorous fashion.

With regards to the contemporary recovery from the global economic recession, we expect those gateway regions that possess a related variety of both advanced maritime services and more traditional gateway functions to recover first. More specifically, it will be those gateways that possess what Hesse (2010) regards as an appropriate assemblage of both "site" and "situation" that will be best positioned in the value chains associated with the global flows of commodities. This conclusion, like so many in the field of economic geography, suggests that policy makers should not expect easy and quick successes as they attempt to upgrade freight gateways into multi-functional advanced service hubs. Nonetheless, the grounded case of Vancouver suggests that the location of head offices and research and development functions of lead firms in the transport sector can contribute to the local upgrading of AMPS from purely routine service functions to regional management and client liaison functions, with potentially higher-value spinoffs. Maritime law especially remains an important locally embedded function, corresponding to the global leading jurisdictions, but with the potential for regional growth.

PART II
Measuring and Improving Gateway and Corridor Performance

Chapter 7
Measuring Port Performance: Lessons from the Waterfront

Claude Comtois and Brian Slack

1. Introduction

Measuring performance has become a necessity in a very large number of fields (see Chapters 8 and 9). From universities seeking to assess teaching, to medical facilities measuring how well they respond to patients' needs, developing procedures for assessment is seen as an essential step to improve performance. Both public and private sectors are moving rapidly to introduce such systems. The transport and logistics industries are particularly appropriate candidates, because supply chains involve a large number of different actors, and the goal of seamless chains requires each segment to operate with high levels of efficiency and fluidity (Gunasekaran and Kobu 2007). It is argued that measurements will enhance ports' supply chain visibility as well as serving as a diagnostic tool to identify operational problems. At the same time measurements can facilitate marketing by demonstrating ports' commitment to improving performance. Finally, it is suggested that performance indicators can help ports with securing funding for infrastructure. These advantages are particularly relevant in the present economic climate where ports are seeking government investment and new clients. Ports are going to be held to higher levels of performance, which will require the collection and dissemination of reliable, quality indicators. Actors who manage to meet these new, rising demands can be expected to gain competitive advantages.

In this paper we explore the issues in developing a set of port performance indicators for bulk transport. As we will demonstrate, the problem of producing a set of metrics appropriate for bulk commodities is challenging because of the diverse nature of the goods, their varied handling methods, the great range in vessel types, and the uneven seasonal patterns of flows in many cases. The paper arises out of a contract with Transport Canada. The Ministry is embarked on a process of measuring port activity with the goal of ensuring transparency, facilitating operational improvements and enhancing fluidity. It has already begun to implement a set of assessments of container operations in major Canadian ports. While the term "performance" is used widely in benchmarking and indicator studies, in applying the selected metrics to container ports Transport Canada has begun to use the term "utilization", since it was clear at the outset that there are too many differences in the individual characteristics and functions of the individual ports to make detailed comparisons possible. Rather, the objective of the exercise is to measure how the metric scores have changed, and hopefully improved, over time. We were mandated to investigate the possibilities of extending the assessments to bulk cargo.

We begin by providing a review of the literature on measuring performance, and reveal the divergence between academic studies and the few practical applications. Based on these studies and interviews with port officials in Canada and Europe we draw up a set of considerations that are necessary prerequisites for any application of Key Port Utilization Indicators (KPUIs). Given the fact that ports are components of logistics chains, it is essential that indicators have utility for stakeholders in those chains as well as the port operators and managers. We follow this up with a review of detailed metrics that could form the basis of a procedure for assessing dry bulk cargoes. Finally, we produce a set of proposals that should guide any actual implementation of a KPUI system for Canada's bulk ports that recognize the concentrated nature of bulk product handling at a few major terminals.

2. Literature Review

The measurement and analysis of port performance has received fairly extensive treatment in a range of contexts. Four main types of approaches are evident. First, academic studies have contributed extensively to questions relating to port efficiency and competition. Second, international agencies have produced a number of approaches to measure port activity. Third, measures have been introduced by governments to assess port activity at a national level. Fourth, there are in-house assessments of performance by certain actors along the maritime supply chain such as ship charterers and ship brokers. The latter are usually confidential and therefore are not considered further here.

2.1 Academic Studies

Academic researchers have had a long standing interest in explaining port performance and assessing the determinants of port competition. Prior to the 1990s, much of this attention was directed at determining the factors that shaped port competition, focusing on issues such as port pricing, port selection criteria, hinterland accessibility, and so on. In many cases the studies were abstract and theoretical, or based on survey instruments and descriptive. Since the 1990s studies have become more quantitative, using such traditional techniques as regression and Principal Components Analysis (PCA), but a new set of quantitative techniques gave economists additional tools to explore port performance and competition. The most common are Data Envelopment Analysis (DEA) and Stochastic Frontier Analysis (SFA) (see Table 7.1).

Table 7.1 Sample of KPPIs Used in Academic Studies (modified from Barros 2006)

Papers	Method	Units	Inputs	Outputs
Roll and Hayuth (1993)	DEA	Hypothetical numerical example of 20 ports	Manpower Capital	Cargo uniformity, Cargo throughput, Service level, Consumer satisfaction, Ship calls
Martinez -Budria *et al.* (1999)	DEA	26 Spanish ports, 1993-1997	Labour expenditure, Depreciation Other expenditures	Total cargo Rental revenue
Tongzon (2001)	DEA	Four Australian and 12 others for 1996	# Cranes # Container berths # Tugs Terminal area Delay time # Employees Book value of …	Cargo throughput Ship working rate
Barros (2003)	DEA	10 Portuguese seaports, 1990-2000	# Employees Asset book value	# Ships Cargo tonnage Break bulk # containers Dry bulk tonnage Liquid bulk tonnage
Park and De (2004)	DEA	11 Korean ports for 1999	Berthing and cargo handling capacity	Cargo tons # Ship calls Revenue Customer satisfaction
Tongzon and Heng (2005)	SFA	25 international ports	Container throughput	Quay length # Gantries Port size Private participation in port
Cullinane *et al.* (2006)	SFA and DEA	28 international ports 1983-1990	Container throughput	Terminal length Terminal area # Quay gantries # Yard gantries # Straddles

Note: DEA is Data Envelopment Analysis; SFA is Stochastic Frontier Analysis.

Data Envelopment Analysis (DEA) is the most widely used method, if the number of citations is considered. Using a set of non-parametric variables, it provides a quantitative measure of the relative efficiency of each case, and identifies which variables account for inefficiencies. A technical problem is that a different set of case studies changes the efficiency frontier and the results for the same ports in a different sample would change. Two practical problems limit the use of these studies. The majority treat the port as a whole unit, and even though they try to eliminate the obvious differences between different types of products by concentrating on container traffic, differences in concessions, management, handling equipment and markets are ignored. Only a few, such as the papers by Roll and Hayuth (1993) and Barros and Athanasiou (2004), consider bulk traffic. A second problem is that very broad and general metrics are employed, such as a measure of container traffic, some indication of port equipment (number of cranes), a metric of physical dimensions (area or length of berths), and labour (number of dockers). Where bulk cargoes are included, tonnage of all bulk handled in the port was used (Roll and Hayuth 1993).

Stochastic Frontier Analysis (SFA) assumes a parametric function, which necessitates strict rules in data collection. This has proven to be a very difficult challenge in many studies, and results in metrics being applied to container traffic for ports in general. Once again most studies focus on containers, an exception being Tovar *et al.* (2007) who consider a broad range of traffic types, including tonnage of dry bulk, liquid bulk and roll-on roll-off grouped together for the port unit as a whole. Recently, Cullinane *et al.* (2006) compared DEA with SFA and found these results were comparable over a common data set.

A wide range of other quantitative techniques have been employed, many which employ much broader metrics. A popular method is Principal Components Analysis (PCA), which provides a means of collapsing large data sets into a smaller set of factors. Examples include De and Ghosh (2003), which is based on metrics collected for the port as a whole, Pando *et al.* (2005) and Sanchez *et al.* (2003) which include data obtained from questionnaire surveys. Once again the emphasis is on container traffic.

Academic papers that have opted for non-econometric approaches have provided some useful insights. The need for benchmarking has been noted by a number of authors. An early paper by Lawrence *et al.* (1997) described the work of the Australian Bureau of Industry Economics and its efforts to assess the level of performance of Australian infrastructure, such as electricity, railways, telecommunication and ports. For ports two measures were used: terminal charges for TEU and terminal charges for coal. These were then compared with several international ports to determine how far Australia ports deviated from best practices. Marlow and Casaca (2003) provided a thoughtful paper on performance indicators. They proposed a range of indicators to assess ports' effectiveness based on time, such as ship waiting time, loading/unloading time, dwell time for cargo, and port costs per unit of tonnage. They also discuss the need to measure and assess management and labour performance. None of these proposed indicators were operationalized, however. Bichou and Gray (2004) provide a trenchant warning about the difficulties of measuring performance at several terminals or ports because of the uniqueness of each case and the complexity of the industry. Furthermore, they note that most studies ignore logistics considerations, despite the role of ports in logistics chains. The authors went on to show that from a survey of the use of indicators at 115 port authorities 44 measured performance using financial data, 37 port authorities used throughput indicators, 24 used productivity measures, and the remaining ten port authorities used economic impact or other indicators. Hardly any of the ports surveyed used logistics measures.

Providing some other perspectives is the work of De Langen (2007, 2008a). He outlines some of the difficulties in dealing with overall port performance indicators, arguing that the port is a complex entity, whereas a terminal or even berth provides a more direct indication of actual activity. In the 2007 paper De Langen and the other two authors propose three broad groups of indicators based on the port products: 1) transfer product – throughput, ship waiting time; 2) logistics products – value added in logistics, m^2 logistics space; and 3) manufacturing product – value added and investment in port related industry. In the 2008 presentation at the European Sea Ports Organisation (ESPO) he added the need to include user perceptions, and an extensive list of other metrics including port connectivity, modal split, KWH of sustainable energy produced in the port, financial incentives for sustainable shipping and so on. It is interesting to observe that having discussed difficulties of treating ports as a whole, the indicators proposed are port-based, and that despite observing that throughput data may be due to factors not related to actual port performance, the proposed indictors of port products included cargo throughput.

2.2 International Agencies

A different set of approaches have been developed by international agencies. The United Nations Conference on Trade and Development (UNCTAD) continues to play an important role in maritime training. From the 1970s it has produced a number of training manuals aimed at helping developing countries upgrade the management and operations of ports. In 1983 it developed a manual to try to establish a uniform system of measuring port activity, by collecting statistics in a consistent fashion (UNCTAD 1983a, 1983b). The Supplement (De Monie 1987) provided a section on performance indicators. These were to be collected by berth by commodity. They included overall port traffic and berth throughputs, ship turnaround time, berth productivity, labour (gang) productivity, berth occupancy rates, and idle time.

This was followed by a World Bank Report (Chung 1993) on port indicators. It identified three broad classes: operational performance indicators, asset performance indicators, and financial performance indicators. These are collected at the level of the berth, with operational indicators being vessel average turnaround times, delay times (identify causes), cargo throughput per ship per hour, tons per gang hour, TEUs or bulk per crane, and dwell time. Asset performance is measured by berth throughput per linear quay and berth occupancy rate. Financial performance are derived from the usual financial reports – relating income and profits to total traffic and vessels handled.

Since 2001 ESPO has been collecting benchmarking data from 25 ports to compare performance. Data are collected at the port level on port size, land allocation to different cargo handling, employment, number of gantry cranes, vessel arrivals by type and gross tonnage, tonnage of cargoes (main bulk and general cargo types), numbers of containers, and numbers of passengers. The data are published in tabular form, and provide average values differentiated by size of ports. This is now being extended to include 85 ports to collect data grouped into productivity/utilization measures (quay productivity, terminal area productivity, storage area productivity and crane utilization), and service quality measures (ship arrival and departure time, ship berth occupancy time, ship loading time, truck turnaround time, gate in and out times, and rail departure from schedule times).

2.3 National Studies

In 2000, the United States Maritime Administration (MARAD) published a report on terminal productivity measures prepared by a consultant (USDOT-MARAD 2000). The report considered the full range of cargo types. It proposed two sets of indicators: measures of output, including tons and dollar value, and five input variables, including terminal area, storage area, storage tons (silos), berths, and man-hours. Each of the input variables were seen to be weighted by units of output. No further evidence of its application has been found.

The most successful and comprehensive approach to measuring port performance is from Australia, where the Australian Bureau of Infrastructures, Transport and Regional Economics (BITRE) has played a major role in benchmarking. Arising out of government legislation seeking to reform port industry labour in 1989, mandated reporting of port performance was undertaken. This reporting involved the participation of the stevedoring industry and provided a range of indicators on a quarterly basis for the container trade. The mandated reporting ended in 1992, although there was a lot of opposition from certain sectors, particularly the stevedoring companies that carry out port operations (Hamilton 1999). On the other hand other sectors of the maritime industry, such as shippers, felt it should continue. While BITRE continued its mandatory responsibility of producing indicators of container stevedoring productivity, it was also asked by Parliament to produce 6 monthly indicators on port costs. These legislative requirements led to BITRE developing a consistent set of measures for the container ports which were subsequently published in a quarterly report, *Waterline* (Australia 2008a, 2008b). Over the years this publication has proved to be extremely useful, allowing the various actors in the industry to obtain a reliable and independent source of information. Eleven metrics are collected (Table 7.2).

Recently, BITRE has indicated it is planning to extend the collection of indicators to bulk ports. It is presently obtaining reactions from the industry. The proposal calls for collection of information for iron ore and concentrates, coal and coke, aluminium ore and concentrates, cereals and cereal preparations, cork

Table 7.2 Selected Metrics Used by Australian Bureau of Infrastructures, Transport and Regional Economics (BITRE)

	Indicators	Measurement
1	Containers handled	Number of containers lifted on/off cellular chips
2	Container turnaround time	Average truck turnaround time in a quarter/average number of containers on a truck in a quarter
3	Truck turnaround time	Time between truck entering and exiting port terminal
4	Crane intensity	Allocated crane hours/elapsed labour hours
5	Crane rate	Number of containers/elapsed crane time
6	Elapsed crane time	Total allocated crane hours – operational and non-operational delays
7	Elapsed labour time	Time between labour first boarding the ship and labour last leaving the ship less non-operational delays
8	Vessel working rate	Number of containers handled/elapsed labour time
9	Ship rate	Crane rate x crane intensity
10	Storage density	Number of container x berth metre squared
11	Elapsed berth time	Average time between arrival and departure from berth

Source: Australia-BITRE, 2008.

and wood. These would be collected for ports that handle in excess of 2 million tons of the commodity. The data would be collected quarterly and published twice a year.

2.4 Summary of the Literature Review

Several conclusions can be drawn from the review of the literature. With very few exceptions, the studies have treated ports as single entities. Academic researchers, in particular, have been forced to use this approach because of data availability. Data access is often aggregated because of commercial confidentiality. This ignores the reality of port operations. Different terminals and berths have a wide range of infrastructure conditions, with different managements and operating conditions, so that when grouped together there are likely to be significant differences in activities and performance (Slack 2007).

Most studies have dealt with container traffic (see Chapter 8). Even if container activities are much more homogeneous than most other cargo types, several authors have commented on the difficulties of comparing container ports (De Langen 2007, 2008a). Very few studies have sought to provide measures of bulk traffic. The metrics employed tend to be very general and sometimes inappropriate. The selection seems to be weighted more because of data availability than relevance. For example, many of the indicators employed, such as TEUs, measure output that may not be directly related to individual port/terminal performance and utilization, since TEU traffic volumes may be due to market size, irrespective of performance.

Finally, very few studies are longitudinal, the data and analysis being carried out at one time only. This is one of the weaknesses of many of the econometric analyses, where for example, the efficiency frontiers are drawn from the particular combination of variables at one point in time. Any assessments of utilization and performance need to be collected over time so that changes/improvements can be monitored, and if necessary adjustments made to operations.

3. Developing Indicators of Bulk Cargo Activity

Bulk cargoes represent a major element in international trade, representing more than 7 billion tons in 2006. Petroleum products (oil and natural gas) and the five major dry bulks (iron ore, coal, grains, bauxite/alumina and rock phosphate) account for 36 per cent and 25 per cent respectively of overall tonnage of world seaborne trade (UNCTAD 2007). With the growth of industrial production in East Asia, imports of bulk products from around the world have created a demand for tonnage additions in the world bulk carrier fleet (ISL 2006).

Bulk commodities have a low value/weight (or volume) ratio, implying that the efficiency of land and marine transport has an impact on value added. The focus on port utilisation levels then becomes critical in assessing logistics costs and the competitiveness of bulk commodity movement. The general principles suggest the need to achieve scale economies, to minimize handling costs, to fully integrate vessels, land transport, terminal and storage in a single system and to reduce inventory. A distinction has to be made between dry bulk commodities (granular, lumpy, loose or powdery substance) and liquid bulk (crude and refined oil, liquified natural gas and chemicals). Common application of port utilisation indicators to dry and liquid bulk is not possible as handling techniques put different demands on facilities, the level of automation is distinctive, and hinterland connections are highly contrasted. For this reason we focus on dry bulk commodities.

3.1 General Considerations

Key Port Utilization Indicators (KPUIs) are not measures of inputs, processes alone or even outcomes *per se*. KPUIs are more than simple statistics describing various aspects of the port system. It is possible to compare ports on the volume of throughput, vessel traffic and employment, as is common in academic studies. But these are not totally within the control of the ports. Throughput rates or number of ships are a function of the world economic conditions or demand while number of employees is closely related to a combination of input factors and skills. There are several misapplications of KPUIs. For example, comparing stevedoring crane rates with other national crane rates is fraught with danger when the unmeasured influences are not considered. Another problem is treating capital investment as a measure of port productivity. This confuses business investment and contribution to the financial health of the port with value added or productivity. Some investments do not produce any new value while some value added is produced outside of the port business investment structure.

A KPUI is a metric that indicates the extent to which the port system, a port function or some internal process is performing as it ought to. KPUI development often employs an Input-Process-Outcome (IPO) framework. Inputs represent what port starts with – physical infrastructure, manpower. Process refers to what ports do with the inputs – transport and trade facilitation processes governing physical flows. Outcomes are the throughput, revenue and service effects of the port activities on customers – cost, supply, on time delivery, incomes, etc. The interaction of input characteristics and port structures and processes to create value-added outcomes all take place within an external environment (demographic, economic, social, regional or political factors) which affects ports but are beyond their control or influence.

KPUIs must reflect a port's most fundamental goals or mission. Each port should be held accountable primarily by reference to how it fulfills the mission it has determined for itself. Is the essential goal of the port concerned with servicing exports? Is it an import and/or export distribution centre? A transit port of call? The most important step in KPUI monitoring is to be very clear about the objectives. This permits monitoring to be linked with the metrics selected for benchmarking.

At the same time as a facility seeking to extract the maximum potential from existing infrastructures, it is necessary to understand the role of the main stakeholders in the development of KPUIs. The main stakeholders are shippers, carriers, terminal operators, and land carriers. Any approach to meaningful KPUI must take into account their interests, capacity and willingness to respond in a transparent and accountable decision-making framework. Shippers are concerned with the distribution of freight at an acceptable rate within a desireable timescale. They are involved in the choice of port for loading and unloading dry bulk commodities. For the shipping industry KPUI reflect the availability of berths and their efficient cargo handling and their capacity for providing rapid turnarounds. For the terminal operators their interests in KPUI are often measured in terms of adequate space, efficient handling equipment, and timely arrival and pick up of cargoes in the terminal. For the land carriers the main issues relate to accessing the terminals with minimal delays.

Benchmarking must therefore involve the participation of different organisational units to facilitate the assessment of process and performance. This collaboration is mutually reinforcing and underpins port improvement. Port stakeholders are not as interested in ranking their organisation against others but rather in recognizing the increasing returns the benchmarking process can generate over time, ensuring their participation in the evaluation process, and providing fair representation and utilisation of indicators.

Dissemination of meaningful KPUIs can go a long way toward demonstrating real accountability. Trust and confidence of both private and public actors will be enhanced if port authorities make their missions explicit, measure the extent to which they are achieving their goals, and share data on successes and problems including what is being done to remedy deficiencies. In Australia, KPUIs have been shown to be useful to many groups, including policy makers, industry groups and the media to provide a better understanding of the port operations system.

3.2 KPUI Data Issues for Dry Bulk

The implementation of KPUI for dry bulk requires appropriate data. The literature review and consultations with several experts in the field and interviews with a dozen port managers in Europe and North America identified key problems in drawing materials for assessment:

- Transport flow data are crucial for analysis of KPUI.
- The quality of data varies greatly by port, by mode and by time period.
- Meaningful statistics are collected by private terminal operators who are often reluctant in sharing information.
- Some basic data concerning dry bulk trade are controlled by overseas stakeholders.
- Audits of the data suppliers' information systems are critical to ensure reliability of data.
- The stevedoring industry has a fundamental role in supplying data.
- Confidentiality of data is a prerequisite in developing KPUI.
- Interpretation of KPUI must be supported by sets of guidelines.
- Collection of data, integrity of KPUI process, reporting on KPUI and dissemination of results are best guaranteed through a high level independent agency.

Developing a preliminary list of possible indicators pertaining to the dry bulk sector underlines the complexity of KPUI. There are different types of dry bulk. The handling conditions of dry bulk materials are influenced by a wide range of factors (size, weight, water content, surface adhesion, ease of flow, extent of compaction). There are various types of contractual arrangement used for the shipment of dry bulk. The decision making centre is rarely at the dry bulk port location. Ships and consignment sizes vary enormously. Handling equipment is often custom designed for specific dry bulk commodities in some cases but in others the same equipment is used to handle different bulk cargoes. Port labour can be assigned various tasks on a wide spectrum of dry bulk commodities.

From the literature and interviews a wide range of indicators were assembled. Table 7.3 lists 37 indicators and their possible measures. They have been grouped into five types of functional areas.

Table 7.3 Selected KPUI for Dry Bulk Ports

Port Functional Sector	Eelement	Indicator	Data Needed	Units of Measurement
Transportation Demand	Traffic	Bulk throughput	Throughput	Tonnage
	Waterways	Vessels processed	Number of vessels entering/leaving the port	Vessels
	Highways	Trucks processed	Number of trucks entering/leaving the terminal area	Vehicles
	Railways	Rail car availability	Number of cars available for loading	Cars
Transportation Supply	Terminal	Terminal design capacity	Maximum possible throughput	Tonnage
		Terminal utilisation rate	Actual vs maximum potential throughput by terminal	%
		Terminal operations	Terminal operating hours	hours
		Terminal productivity	Throughput and storage space per terminal	Tonnes per hectare
	Berth	Berth design capacity	Maximum possible throughput by commodity and type of ship	Tonnage
		Berth utilisation rate	Actual vs maximum potential throughput by berth	%
		Berth operations	Berth operating time	Hours
		Berth productivity	Throughput and berth length	Tonnes per metres
	Equipment	Equipment design capacity	Maximum possible throughput by commodity and type of machine	Tonnes
		Equipment utilisation rate	Total availability of machine minus down time	Hours
		Equipment operations	Machine operating time	Hours
		Equipment productivity	Throughput and machine operating time	Tonne/hour/machine
	Vessel	Stock-to-vessel requirements ratio	Stock held in inventory and demand requirements of vessels	Tonnage
		Stock-to-shipment ratio	Stock held in inventory and actual volume loaded	Tonnage
Employment	Labor	Labor generation	Direct and indirect employment generated by port activities	Manpower hours
		Labor income	Annual wage and benefits	$
		Labor productivity	Throughput, working time, income	Tonne/hours/$

Table 7.3 *Continued*

Port Functional Sector	Eelement	Indicator	Data Needed	Units of Measurement
Transport Fluidity	Waterways	Vessel design capacity	Maximum possible throughput in tonnage by commodity and type of ship	Tonnage per loaded vessel
		Vessel loading utilisation rate	Actual loading per vessel vs maximum potential loading capacity	% and standard deviation
		Vessel loading time	Variation in loading time	% and standard deviation
		Vessel operations at port	Vessel start and end date time in port area	Vessel-hours
		Vessel operations at berth	Vessel start and end date time at berth	Hours
		Vessel schedule reliability	Vessel scheduled arrival date time vs vessel actual arrival date time	% and standard deviation
	Highways	Truck booking system	Number of truck booking in the system and number of trucks arriving at the terminal	% and standard deviation
		Truck turnaround time	Truck arrival and departure date time	Hours
		Gate congestion	Minimum and maximum waiting time of trucks at the gate	Minutes
	Railways	Rail loading utilisation rate	Actual throughput vs maximum car capacity	% and standard deviation
		Rail loading time	Variation in loading time	% and standard deviation
		Train cycle time	Time train takes to make trip from source back to source	Hours or days
		Rail operations at terminal	Rail start and end date time at terminal	Hours
		Rail schedule reliability	Train scheduled arrival date time vs train actual arrival date time	% and standard deviation
		Train space occupation	Length of cars and throughput handled	Metres per tonnage/ train length
Environment	Emissions	Air emission	Throughput and air contaminants	Particulate matter (PM) concentration/tonnes

Transportation demand indicators focus on the volume of dry bulk processed and handled through the port and on traffic that flows through the port in terms of vessel calls at the port, truck movement or rail activities.

The usefulness of transportation supply indicators depends on plans elaborated and targets that have been set. Indicators must have a purpose: to remedy a problem? to fix costs? to set prices? Equipment productivity needs to meet industry targets. Such indicators provide a diagnostic on productivity and a link between product and process to technological innovation. The best assessment is the level of productivity that can be maintained on the long term. As a result the targets are best met using a working time sheet.

Labour skill, attitude or motivation are encompassed in the general productivity indicators. Employment figures from the operation and management of the dry bulk industry sector need to be considered as indicators for their multiplier effect influencing the expenditures of the port, the supplier industries and the economy in general.

The logistical system structure of a dry bulk port comprises a waterside component (berthing/unberthing activities) and a landside component (rail or road carriage). Tonnage is not important as it depends on customer's demand. What is required is to establish a service quality index. Even if the schedule is not under the port's direct control, the objective is to measure how many times the port met expected time of arrival and departure for ships, road and/or rail. The scale ranges from 0 per cent when the port is closed to 100 per cent when all shipments are handled on time.

Environmental indicators have become an unavoidable part of benchmarking process in the marine industry. Climate change and greenhouse gas emissions are a reality. But no standards on measuring polluting emissions have been agreed upon. Our research suggests that initially the salient aspects of air emissions as measured by the volume of particulate matters generated by the port activities should be considered as a base for adding further indicators at a later date.

KPUI monitoring is one of several quality enhancing activities designed to improve quality and accountability for bulk ports. The indicators employed should meet certain standards notably scientific rigor in terms of feasibility, robustness, usability, transparency and readability:

- Feasibility involves issue to the extent to which the data needed for constructing the indicator exist.
- Robustness is the quality of the indicator to withstand changes in operational environment or conditions without affecting its functionality to overall KPUI.
- Usability denotes the ease with which port stakeholders can easily employ KPUI with low rate of error.
- Transparency pertains to the level a customer is involved in the decision making process of KPUI accepting goals and operations of indicators.
- Readability refers to the intelligibility of an indicator and the ease with it is understood by a stakeholder.

In Table 7.4 we develop a set of criteria to measure these guidelines, which are then applied to the suggested bulk indicators (see Table 7.5). While there is some element of subjectivity, the table provides an indication of the challenges that must be met in developing a useful KPUI system.

Table 7.4 KPUI Standards, Suggested Weightings

Criteria	Rank value	Description
Feasibility	1	Poorly organized data collection
	2	Some basic data quality controls are in use
	3	Data is quality-controlled and routinely available for operations and to stakeholders on request
	4	KPUI's fully transparent data management system provides high-quality information on all aspects of the bulk port to all interested parties
Robustness	1	Indicator is used but without flexibility to respond to changes
	2	Port stakeholders respond to operational changes without links to KPUI measurement
	3	The port stakeholders targets KPUI to enhance bulk port organization
	4	The port stakeholders are committed to maximizing and continuous improvement of KPUI benefits through efficient use of resources
Usability	1	There are no adequate system for defining KPUI procedures
	2	KPUI procedures are documented but remain ineffective
	3	KPUI procedures exist and appropriate training available without feedback mechanisms
	4	Effective KPUI system in operation, integrating best practices with responsive decision making and commitment for continuous improvement
Transparency	1	No communication with customer
	2	Customer information available on request
	3	Opportunities exist for customer to voice their view
	4	Customers are encouraged to contribute to KPUI planning and decisions
Readability	1	There is no agreement on a Canadian national structure of KPUI
	2	Evidence of institutional arrangements for KPUI defining terms and specifying roles and responsibilities of CPA and different stakeholders
	3	Definitions and responsibilities of KPUI are clearly specified
	4	Clear and rigourous set of underlying terms, well-established KPUI system and evidence of an audit control

Table 7.5 Suggested KPUI Standards Criteria Applied to Bulk Indicators

Port Functional Sector	Indicator	Criteria				
		Feasibility	*Robustness*	*Usability*	*Transparency*	*Readability*
Transportation Demand	Bulk throughput	3	3	3	3	3
	Vessels processed	3	3	3	3	3
	Trucks processed	2	3	3	3	3
	Rail car availability	3	3	3	3	3
Transportation Supply	Terminal design capacity	3	2	3	3	3
	Terminal utilisation rate	3	2	2	3	3
	Terminal operations	3	3	3	3	3
	Terminal productivity	3	2	3	3	3
	Berth design capacity	3	2	3	3	3
	Berth utilisation rate	3	2	2	3	3
	Berth operations	3	2	3	3	3
	Berth productivity	3	2	3	3	3
	Equipment design capacity	3	2	2	3	3
	Equipment utilisation rate	2	2	2	3	3
	Equipment operations	2	2	2	3	3
	Equipment productivity	2	2	2	3	3
	Stock-to-vessel requirements ratio	3	3	3	3	3
	Stock-to-shipment ratio	3	3	3	3	3
Employment	Labor generation	1	1	1	2	1
	Labor income	2	1	1	2	1
	Labor productivity	1	1	1	2	1

Table 7.5 Continued

Port Functional Sector	Indicator			Criteria		
		Feasibility	*Robustness*	*Usability*	*Transparency*	*Readability*
Transportation Demand	Bulk throughput	3	3	3	3	3
	Vessels processed	3	3	3	3	3
	Trucks processed	2	3	3	3	3
	Rail car availability	3	3	3	3	3
Transport Fluidity	Vessel design capacity	3	2	3	3	1
	Vessel loading utilisation rate	3	2	3	3	1
	Vessel loading time	3	2	3	3	1
	Vessel operations at port	3	2	3	3	1
	Vessel operations at berth	3	2	3	3	1
	Vessel schedule reliability	2	1	1	3	1
	Truck booking system	1	1	1	1	1
	Truck turnaround time	1	1	1	1	1
	Gate congestion	1	1	1	1	1
	Rail loading utilisation rate	2	1	1	2	1
	Rail loading time	2	2	1	2	1
	Train cycle time	2	1	1	2	1
	Rail operations at terminal	3	2	1	2	1
	Rail schedule reliability	2	1	1	2	1
	Train space occupation	3	1	1	2	1
Environment	Air emission	2	2	1	3	1

4. Applying KPUIs to Canada's Dry Bulk Ports

Transport Canada has indicated it wishes to develop a KPUI procedure for the 17 Canada Port Authorities, the country's largest public ports. We suggest that KPUIs for dry bulk should be measured only for key commodities and at ports within which these cargoes represent a dominant function. This suggests the adoption of selection criteria by port and by market segment. An example of the priorities for implementing KPUIs based on the demand profile of minerals and metals for each of Canada's Port Authorities is contained in Table 7.6.

High priority is awarded to CPAs handling above 25 per cent of Canada's yearly throughput for a specific dry bulk. Medium priority is granted to ports handling between 10 per cent and 25 per cent of Canadian trade for given product during a year. Low priority is given to ports handling between 2 per cent and 10

Table 7.6 KPUI Priority Levels for Metal and Mineral Market at Canada Port Authorities

	Coal	Gypsum	Iron Ore	Non-ferrous metal	Other metallic ores	Other non-metallic ore	Potash	Sulfur
Belledune	-	-	--	--	-	-	--	--
Chicoutimi (Port Saguenay)	--	--	--	--	--	--	--	--
Halifax	--	++	--	-	--	-	--	--
Hamilton	-	-	+	--	--	--	--	--
Montreal/ Contrecoeur	--	-	-	+	+	+	--	--
Nanaimo	--	--	--	--	--	-	--	--
Port Alberni	--	--	--	--	--	--	--	--
Prince Rupert	-	--	--	--	--	--	--	--
Quebec/Lévis	--	--	-	-	++	--	--	--
Saint John	--	--	--	--	--	--	+	--
Sept Îles/Pointe Noire	--	--	++	++	--	+	--	--
St. John's	--	--	--	--	--	--	--	--
Thunder Bay	-	--	--	--	--	--	-	--
Toronto	--	--	--	--	--	--	--	--
Trois Rivières	--	-	--	+	--	++	--	--
Vancouver/ Fraser River	++	--	--	+	++	-	++	++
Windsor	--	--	--	--	--	-	--	--

Note: -- not significant/- low priority/+ medium priority/++ high priority. Classification is based on CPA traffic data for 2006 provided by Transport Canada.

per cent of Canada's traffic for a commodity. CPA handling less than 2 per cent of a given bulk in a year is not considered significant. This classification is not an assessment of the local importance of a given traffic. Rather the objective is to identify thresholds to justify the development of relevant KPUI in the context of minimising the cost of data collection.

The KPUI framework applicable to Canada's bulk ports should be regarded as a guide rather than a strict scheme of prescriptions. The framework has not been worked in detail as the main objective has been to demonstrate the approach. As the main purpose of KPUIs is about communicating knowledge and advice and raising awareness among port stakeholders, it is suggested that a pilot project among selected Canada's bulk ports be conducted. Before implementing a nation wide approach, testing port specific measurements is needed. The suggested CPAs from which such pilot tests could be carried out include Sept-Îles, Québec, Trois-Rivières, Montréal, Thunder Bay, Vancouver and Prince Rupert.

5. Conclusion

Conceiving and implementing a system of KPUIs has the potential to be a contentious and complex issue. The objective of any framework should be to help increase public awareness and enhance the relevance for port stakeholders. The framework should open up discussions about the relevance, limits and strength of indicators. This should facilitate monitoring strategies. Application of KPUIs at the individual port or national level does not mean that the port itself or country's port system lags behind the inexorable evolution of port efficiency. KPUIs should be regarded as a logical but essential management tool.

Bulk shipments provides an important marker on the commercial geography of Canada and reactions and response to major trends in the global economy. Opportunities and challenges for Canada's market position in global dry bulk trade will depend in part on the competitive advantage of Canada Port Authorities. Canada's dry bulk trade growth will increasingly be inscribed within global logistics networks. Most of the principal actors, with a few notable exceptions, will be international. To retain and expand the Canadian segments of the chain will necessitate increasing attention to KPUIs. Future scenarios will have to consider the provision of KPUIs in order to increase market share, assess site and equipment productivity, strengthen infrastructure capacity, evaluate innovations penetration, monitor sustainability, improve fluidity and display transparency.

Chapter 8
Key Interactions and Value Drivers towards Port Users' Satisfaction

Athanasios A. Pallis and Thomas K. Vitsounis

1. Introduction

Maritime-based freight transportation has been affected by major changes in trade systems. The quest for port integration in supply chains results in the functional and spatial expansion of port-related activities ("port regionalization" – Notteboom and Rodrigue 2005; Ferrari *et al.* 2006). International ports aim at attracting large-scale companies, drawing extra investments, exploiting resources and attaining efficiencies. Yet the users' complaints arise even when efficiency is achieved (Farrell 2009). Effectiveness is a performance component of increased importance (Brooks and Pallis 2008). Ports need to rethink the measurement of their performance and systematically monitor whether they serve their users both efficiently and effectively in order to fully understand the users' needs (see Chapter 9). They can then outperform competitors via "capture value" strategies that address unmet needs via discovery and exploitation of relevant opportunities (Magala 2008).

While it is acknowledged that the users' evaluations contribute to a comprehensive port performance assessment, it still remains unclear how these evaluations can be methodologically framed and measured. This shortcoming lies partly in the difficulty produced by the complexity, diversity and extended scope of the relations between port service providers and users. Relevant interactions include networks wherein a firm's performance has an impact on its clients' performance and so forth. It also lies in the fallacy that aggregate flows across a network are homogeneous and actors confront them *sui generis*. Empirical evidence has established, however, that strategic choices of involved firms shape these networks, with these choices based on, or highly influenced by, the relations that develop with other actors, calling for actor-centered studies that pay attention to these relations (Hall 2004).

Port users refer to choices, value drivers, and satisfaction thinking of ports as "elements of value driven supply chains" (Robinson 2002). A freight shipper stated during an interview conducted for the present study, "A port always stands within a network where the inland part is as important as the sea part [of transportation]. We always look into the whole supply chain". Still, considering them as any other link would be misleading. The users of maritime-related chains cannot substitute or bypass the port, while the added value to the total transportation path is remarkable. The influence and control that ports exert throughout the supply chain justify a detailed port-specific analysis.

At the same time, port interactions constitute only a fraction of supply chain interactions. As the broader chains, ports consist of multiple layers of involved entities, pursuing their own goals and strategies. In many instances, these entities might be unrelated to the level of value that the port, and/or the chain offer as a whole. These entities are frequently both competitive and cooperating partners, involved in complex internal relationships and wider business portfolios that blur the classification of pairs of market players as competitors or partners (Lamming 2002). The detailed analysis of port-specific interactions does not imply a "black box" logic but attempts to focus on the "part" and better understand the fracture.

This chapter discusses a qualitative field research involving semi-structured interviews with stakeholders in four international European ports (Antwerp, Piraeus, Thessaloniki, Zeebrugge). This sample provides evidence from ports of varied organizational and operational characteristic structures that serve diverse markets. The aim is to reveal two major issues:

1. The structures of the key commercial and operational interactions between port users, service providers, and port authorities – i.e. which are the stakeholders involved, in what type(s) of interaction(s), etc.
2. Specific value drivers towards port users satisfaction.

The research is "cargo-based", with containers being the type of cargo under examination. The fact that actors' strategies in each trade are different calls for attention to the unique set of actors and partners of ports in a particular trade (Hall 2004). Containerized trade in particular implies organizational changes leading to a remarkable focus of port studies on this type of trade (Pallis *et al.* 2010), and the present study contributes to this debate.

Before the presentation of the results, Part II develops a conceptual framework on value measurement and users' satisfaction and establishes how this framework contributes to a balanced Business Performance Measurement (BPM). The concluding section provides an overall assessment.

2. Interactions and Value Drivers

The majority of port performance approaches are directed towards the measurement of productivity and efficiency via the use of internally (i.e. operational) generated information (Gonzalez and Trujillo 2009). Some recent studies attempt a Business Performance Measurement (BPM) that incorporates externally generated information (Guardado *et al.* 2004; Su *et al.* 2003), as do studies of port attractiveness (Ng 2006) and choice criteria (Lirn *et al.* 2003; Tongzon 2008). The users' perspectives remain a key but under-researched performance component.

BPM also moves from pure financial paradigms towards more balanced ones, measuring non-financial information and external success in juxtaposition to internal performance (Bourne *et al.* 2000; Kennerly and Neely 2002). The users'/ customers' perspectives are assessed into the majority of multi-dimensional BPM frameworks, as businesses compete in an environment where value, not price, is the key driver (Neely 1999). In this context, "users" satisfaction' measurements have been popular (Anderson and Fornell 2000).

In port studies, such a tool should look into the specifics of a port as a regionalized system, rather than focusing on individual entities (terminal operators, warehouses, rail, trucks etc.) only. Via their interaction, these actors develop relations falling within a business-to-business (B-2-B) framework. B-2-B relationships are frequently long-term, close and involve a complex pattern of interactions between and within each company (Ford 1980). When research in consumer goods typically relates satisfaction to a single discrete transaction (transactional specific), for B-2-B relations, satisfaction is also an output of the relations that the two involved parties have developed (relationship specific).

Grounded on the differences between industrial and consumer markets, Gross (1997) suggests a replacement of the "satisfaction construct" by the "value construct" as a better predictor of outcome variables. Value is the perceived preference for and evaluation of those product attributes, attribute performances, and consequences arising from use that facilitate or block, achieving the customer's goals and purposes in use situations (Woodruff 1997). Users receive functional and relationship value. Hybrid forms of relationships might also exist. Value identifies which service dimensions are central to a user's ability to attain desired end conditions, and how these dimensions relate to each other and to the user.

Satisfaction judgments complement a value hierarchy by providing feedback on reactions to value received (actual versus standard performance – usually "expectations") and information on the users' feelings about the value received (i.e. how well an organization's value creation efforts are aligned with users' requirements). In general, satisfaction is an effective construct whereas perceived value is a cognitive variable. In recent years, the academic debates the impact that the two constructs have upon behavioural intentions such as word-of-mouth, repurchase intention and willingness to search for alternatives.

Applying the value construct in ports implies a "B-2-B customer value hierarchy" (Woodruff and Flint 2003). Value is achieved by tangible and operational aspects, as well as by intangible and managerial/entrepreneurial ones. A series of actions take place within contextual settings and produce value for port user(s) (Barber 2008). These actions result in relations between the ports' actors that are not static or dyadic. A chain of industrial relations exists, where neither the "end user" nor his actual importance are easily recognized by all the parts of the chain. In addition, every supplier has an impact upon his customer's customer and so on. Therefore, there are three levels of interactions (contexts) to be taken into account in a value measurement: (a) the inner context whereas two firms set up a focal network relationship; (b) the connected network whereas a number of connected business relationships, such as the customer's customers, other suppliers or other customers, other ancillary firms etc., exists; and (c) the outer context of the connected network which is an extension of the connected network directly relevant to a customer-supplier relationship and its inner context (Tikkanen and Alajoutsijarvi 2002).

A number of questions need to be answered towards constructing an overall assessment of port users' value generators. Contemporary developments transform traditional relationships between services providers, users and PAs into new complex ones. The distinction between competitors, clients, and partners frequently blurs. A fundamental issue relates to the nature of the interactions between the market actors. With ports being "clusters of economic transactions" (De Langen 2004), the identification of the actual interactions (relation, or transaction specific episodes) and the different (actual or perceived) consequences that are placed by the participating stakeholders upon each of them (do they act as drivers towards satisfaction?) are prerequisites towards this assessment.

3. Key Interactions in Container Ports

The exploratory study conducted in the four international European ports generates knowledge on (a) how port users, service providers and authorities interact within a port (i.e. level and extent of interactions) and (b) the most crucial interactions and potential other key elements that are, or perceived to be, important for value generation. The interviewed stakeholders were selected on the basis of experience, variety of characteristics and willingness to participate.

The "confidentiality clause" involved in some of these interviews poses restrictions in the ways that the qualitative research findings are presented. Codenames are used in the forthcoming analysis for the identification of each stakeholder. The list includes port authorities (PAs), international terminal operating companies (TOC), freight forwarders (FF), shippers (SH), and liner shipping companies (SL). When applicable, a distinction is made between global shipping companies (GSL) and those operating at a local/regional scale (RSL). Reference is also made to operators (ITO), dockworkers professional organizations (DPO), depots (DEPOT) and inland terminal operators (IT).

3.1 Operational and Commercial Interactions

Interviewees confirmed the benefits of endorsing an "interaction approach", and revealed three key elements that affect the level of value that stakeholders extract from a port:

a. The operational interactions between port users and service providers;
b. The commercial interactions between port users and service providers; and
c. The structural characteristics of the port – geographical location, socioeconomic conditions, etc.

The conducted research also revealed the presence of two distinctive port settings. In both cases, the total of the interactions that take place are operational and commercial ones. Operational interactions involve the physical interaction between two actors on an operational level. That includes the actual delivery of a service (for example the loading of a container into a truck). Commercial interactions refer to all other interactions that are developed between the two parties such as communications, payment, agreements etc., and are grounded more on developed relations rather than physical transactions.

Figures 8.1-8.4 provide representations of the different entities that interact in each setting, either on operational level (Figures 8.1 and 8.3) or on a commercial level (Figures 8.2 and 8.4). Those companies that are in dotted boxes are either subsidiaries of a shipping company or involved in a joint venture with one of them, thus these parties develop coordinated management and strategies. Their operational interactions, denoted with double arrows in the four figures, might be regarded as "internal" operations of limited importance and influence upon the issues of value and satisfaction generated by third party activities'. The dotted arrows denote indirect interactions between two entities.

3.2 The Common user Terminal Setting

The first setting is characterized by the presence of a global shipping line (GSL.1) while its subsidiary (SL.1) offers logistical services by chartering means of transport. SL.1 might also possess and operate a number of barges, trains, or trucks. Between the shipping line and the international terminal operator (ITO.1) a hybrid relation is developed. The global shipping line might be involved in financing the development of the terminal operated by the international terminal operator but enjoys no special treatment (rates, priorities and so on). Rather than that, it might have the obligation to use the specific terminal when it reaches the specific port. Freight Forwarding (FF.1) and regional shipping companies (RSL.1) offer feeder services. The global shipping line does not charter regional shipping companies, but operates autonomously, e.g. a shipper charters them via freight forwarders or, less frequently, directly. For reasons of simplicity, individual shippers are not presented in the figures and freight forwarders act as a proxy for this users' group. An additional presence is the inland terminal, operated either by the global shipping line (IT.1) or by another private entity (IT.2). Emerging shipping lines' logistics subsidiaries, retain long-term agreements with inland terminal operators, in order to consolidate containers in hinterland positions. Those companies that operate them also provide container transportation services. Alternatively, or simultaneously, freight forwarders and the shipping lines' logistics subsidiaries might employ their own means of transportation.

Figure 8.1 describes the operational interactions that take part in the common user terminal setting. Global shipping lines use navigation services provided by the PA (i.e. the harbour office) until the vessel is safety moored. This involves the provision of tugs, pilots and locks (if any). For reasons of grouping, customs are included in this type of interaction, though this is not a classic "deep-sea" procedure. Global shipping lines also maintain indirect (denoted with dotted arrows) operational interactions with freight forwarders and any shipping lines' logistics subsidiaries, as these two players receive and forward containers transported by the first one. Their interactions are indirect as the stevedoring company always intervenes. Terminal operators use services from local dockworkers professional organizations (DPO), as commonly obliged by the port rules. A terminal operator provides stevedoring services to the shipping company and delivers the containers to either (a) forwarders, whether GSL subsidiaries or not, in order to direct it to the hinterland; (b) regional shipping lines, to feed another port; or (c) inland terminals operated by a shipping line or other private ones. Logistics companies and freight forwarders might also use inland terminals for container storage until the organization of further transportation services. Assuming that the PA is not involved in cargo handling, operational interactions are developed between the PA and all port users via the controlled by the PA EDI systems.

The commercial interactions portrayed in Figure 8.2 supplement the operational ones. On a commercial level, global shipping lines interact with the PA (i.e. port dues), develop regular commercial interactions with the terminal operator, for the supply of stevedoring services, and usually with inland terminals, in order to use storage areas in the hinterland. These are frequently sealed by long-term agreements. Additionally, global shipping lines commercially interact directly with the freight forwarder that buys services for container transportation. Logistics companies and inland terminal operators, operated by the shipping line or by a private one, sustain commercial interactions with global shipping lines. When these companies are GSL subsidiaries, these interactions are "internal" and their volatility diminishes. A terminal operator maintains commercial agreements with both the PA, typically leasing the terminal area, and organizations acting as employment pools. Regional shipping lines maintain commercial interactions with terminal operators, i.e. payment for using stevedoring services, and freight forwarders or shippers for feeder services.

The relations between global shipping lines and freight forwarders and inland terminals operated by privates are of a hybrid nature. These entities maintain commercial but not operational interactions, as a terminal operator carries out the operational part. A detailed discussion of these hybrid forms of relations is found in section 3.4.

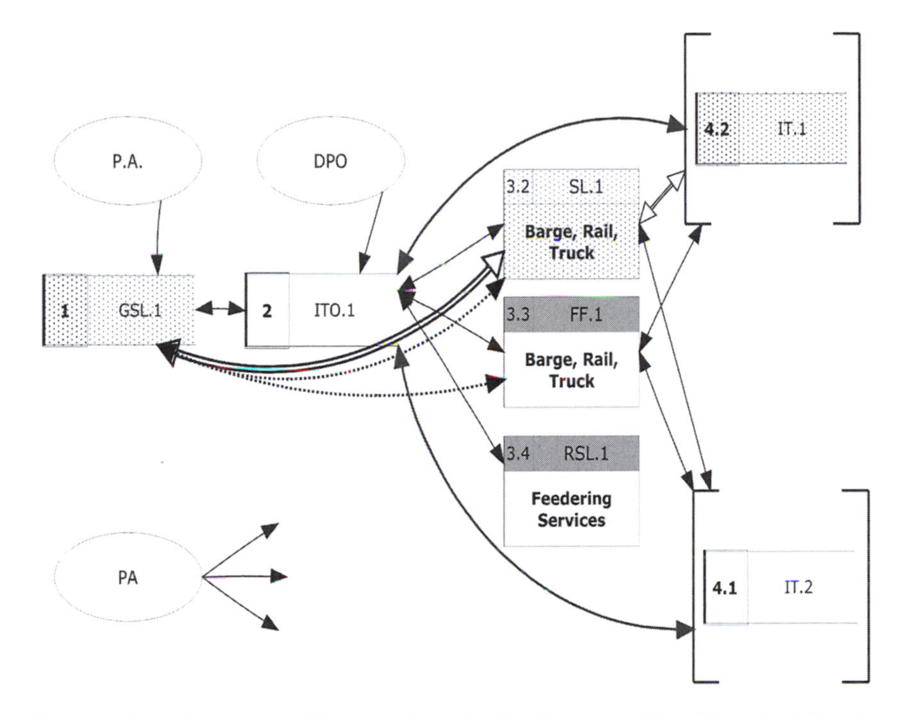

Figure 8.1 Operational Interactions in the Common User Terminal Setting

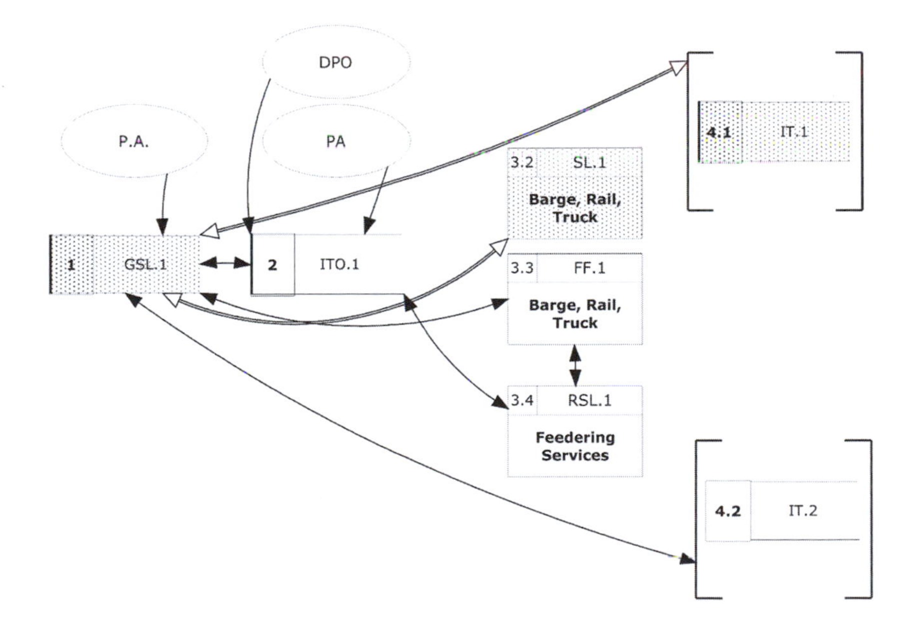

Figure 8.2 Commercial Interactions in the Common User Terminal Setting

3.3 The Dedicated Terminal Setting

The second setting is characterized by the presence of a dedicated terminal. A global shipping liner (GSL.2) typically participates in a joint venture with an international terminal operator (ITO.2) that develops and operates a dedicated terminal. Operational responsibility lies with the terminal operator, whereas the global shipping line uses the specific port as a main hub for the region where it consolidates large volumes of containers. A sister shipping line (SL.2) offers logistical services using rail, barges and trucks that are chartered by third parties. Freight forwarding companies (FF.1) and the responsible PA are also present. The global shipping liner charters regional shipping lines (RSL.2) in order to assist in maritime feeder services whenever the capacity of GSL.2 owned vessels is not sufficient. Depots owned and operated by the global shipping liner (DEPOT.1) or an independent private entity (DEPOT.2) are also part of the setting.

Figure 8.3 describes the operational interactions in this setting. The PA provides services used by the global shipping liner, which maintains indirect (denoted with dotted arrows) operational interactions with freight forwarders, and any regional shipping lines that receive and forward containers transported by the global liner. The terminal operator uses the services of the local dockworkers' organizations, provides stevedoring services, and delivers containers to (a) a sister shipping line and freight forwarders, or to (b) the global shipping liner and the regional shipping lines for maritime feeder. Containers might be

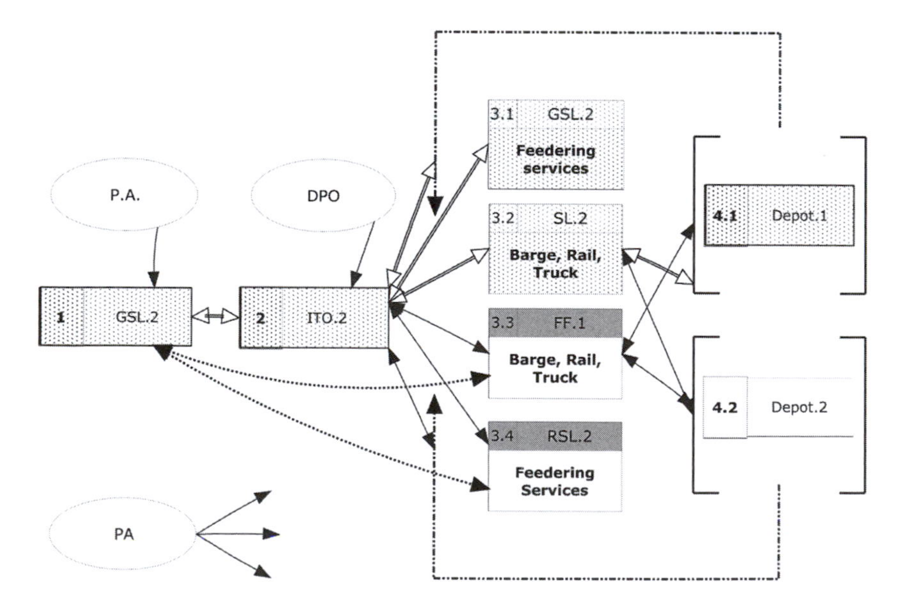

Figure 8.3 Operational Interactions in the Dedicated Terminal Setting

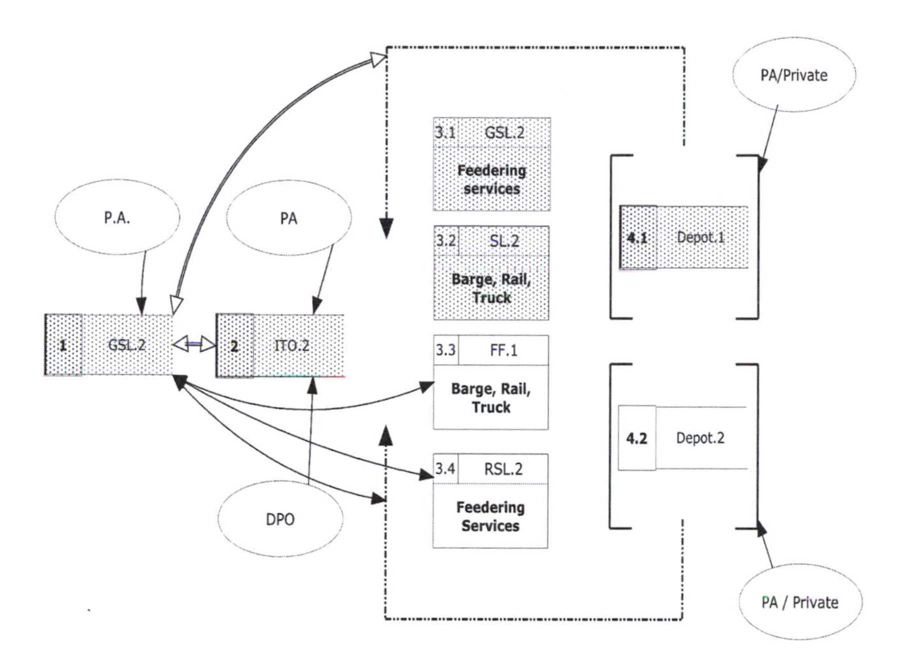

Figure 8.4 Commercial Interactions in the Dedicated Terminal Setting

stored before being forwarded and depot-services are supplied right after the containers' unloading at the terminal (denoted by dotted and dashed arrows). The terminal operator also provides services to depots that are responsible for the terminal-depot movement of containers. In this case, depots interact with the shipping lines and forwarders by releasing the stored containers to them. Operational interactions are developed between the PA and all port users via the PA controlled EDI systems.

Figure 8.4 presents the commercial interactions that take place. The global shipping line (GSL.2) interacts with the PA, and maintains commercial interactions with the terminal operator. Yet, these interactions are "internal" in nature. The same applies to the commercial relations it maintains with involved sister shipping lines and depots operated by subsidiaries. Whenever a private entity operates the depot, it is most probable that the GSL.2 retains long-term direct commercial agreements with it for the storage of containers, while it develops direct commercial interactions with freight forwarders and any regional shipping line chartered for providing feeder services. The terminal operator develops commercial interactions with the PA, commonly a terminal concession, and any organizations acting as employment pools. Finally, depots are involved in commercial interactions with the owner of the land they lease, whether the PA or a private entity.

An internal transportation channel, consisting of companies marked in dotted boxes in Figures 8.3 and 8.4, is present in this setting. This channel enables the involved companies to downgrade the level of influence the connected networks have on them. This generates a process that might transport a container to its final destination without external interactions. The impact of any "not in the channel" company on freight transportation, and consequently on the level of value generation by it, is eliminated.

Corresponding to the conceptual discussion in section 2, the configuration is static in none of the two settings. There is a flow of satisfaction streaming throughout an end-to-end path of a port chain. To give an example, the value that the global shipping lines exert is not only limited to the services offered by the terminal operator but reaches the end of the chain albeit evaporated. Eventually, the (dis)satisfaction of a company is the outcome of its direct supplier activities, and to a lesser extent of its supplier's suppliers and so on. To give an example, a freight forwarder might enjoy high levels of value, but this does not necessarily mean that directly interacting entities produce it. Value might be produced by a distant entity, like the shipping company, the actions of which spill over the chain.

3.4 Hybrid Forms of Interactions (and Relations)

In both cases a number of hybrid relations develop between the entities involved. A hybrid relation develops when a port user maintains a commercial interaction with a specific entity but a third party carries out the operational part of that agreement. These relations, which are infrequent in other industries, add to the complexity of the port settings and any measurement tools based on the stakeholders' interactions.

The most common example is the interaction between a freight forwarder (FF), a shipping line, and a terminal operator. The freight forwarder maintains a commercial agreement with the shipping line that governs issues like rates, number of containers that are transferred, the port of delivery/arrival, and every other aspect of the seagoing container transportation. Communication channels are developed to ensure constant exchange of information throughout the various stages of the process. Still, these entities are not involved in any direct operational interaction. The operational part of the agreement is executed through the terminal operator, without the freight forwarder having any type of agreement with it. Rather than that, the FF delivers the containers to the terminal operator in order to be forwarded to the shipping line without any choice regarding which terminal operator the shipping line is going to use. Thus, the freight forwarder does not retain any bargaining power, i.e. quality, or price demands, *vis-à-vis* the terminal operators.

From the freight forwarders' viewpoint, further complexity exists in terms of value, as he seeks the fulfilment of specific demands by the terminal operator. For example, the actual in-terminal waiting time for the delivery of a container time and its reliability might be the causes for shifting to another shipping company, even though dissatisfaction is caused not by the initially used shipping company.

A second hybrid relation was identified in the common user setting, with the two involved actors (that is a shipping line and a freight forwarder) developing a port service user–provider relationship but being competitors at the same time. The shipping line directly competes with its customers – that is, the freight forwarder, for the movement of containers through its sister company that offers inland transportation services. Any shipping line's sister company will or at least is expected to enjoy a preferential treatment comparing to an independent third party.

A third commonly observed hybrid relation takes place when the global shipping line finances the company operating a concession of a common user terminal and at the same time acts as a user of that terminal. In terms of value measurement, there is no further complexity when the financing of the terminal stands as a distinctive business activity. However, the specific financial activity typically goes hand-in-hand with additional obligations for the shipping company, like the use the specific terminal if one of its ships reaches the port.

4. Value Drivers

Table 8.1 summarizes the revealed value drivers for each group of port users. A general but remarkable finding is the diversification of stakeholders' perspectives as put forward by different port users quoting different elements when asked "What creates value for you as a port user?" The mentioned elements have been conflicting with even the same type of port users referring to different value generators, depending on the setting variations or their particular position within the specific network created in the setting. This is not to say that there are not groups of users sharing the same thoughts, or identifying common value drivers. In essence, the findings support the view that different settings and objectives result in different port strategy-structure-environment matching frameworks (Baltazar and Brooks 2007).

Research also confirmed that not all relationships developed are of the same value. Theories developed and tested in one relational context may not be equally valid in the context of a different type of relationship. Precisely as established elsewhere (Hausman 2001), not all inter-organizational relationships need to be strong ones and efforts to move along the continuum towards a stronger relationship might be wasteful in certain instances. Moreover, what constitutes value appears to be highly idiosyncratic and may vary widely from one user to another; user value is something perceived by the user himself rather than objectively determined by the supplier (Hu *et al.* 2009).

4.1 Shipping Lines

In the case that shipping companies visit common user terminal settings, top value drivers for the shipping line include (a) the existing combination of port

Table 8.1 Drivers Towards Port Users' Satisfaction

	Shipping company with dedicated terminal	Shipping company with no dedicated terminal	Feeder shipping company	FF	Shippers
Port setting	• Geographical location of the port • Social, economic and political conditions • Connectivity with hinterland – Multimodality • Quality of services • Productivity	• Combination of cost and productivity • Multimodality • Quality of services • Available capacity of hinterland connections • Captive market • Labour (skilled and flexible)	• Availability of cargo • High rates • Cost of services • Time spend in a port • Guarantee of entrance	• Absence of congestion in sea and land • Proximity with the market • Productivity • Cost • Smooth customs clearance • Multimodality • Large number of different stevedores	• No delays • Accuracy • Sequence of port of call in a loop • Cost
Shipping lines				• Reliability • Space commitments • Cost • Transit time • Direct call (no transhipment) • Equilibrium between vessel size and number of ports of call	• Lead times • Sequence of port of call in a loop • Days of arrival and departure
Terminal Operator	• Adaptation to their needs • Experience • Confidence	• Combination of cost and productivity	• Guarantee of entrance • Time spent in a port	• Productivity • Absence of congestion in sea and land	
Port Authority	• Marketing and promotion (image improvement) aiming to attract larger volumes of cargo, • Active role in problems resolving, • Availability and exchange of information, • Coordination of port activities ("chain approach").				

productivity and costs; (b) multimodality, including availability, quality of existing links and capacity for further hinterland connections; and (c) the presence of a local captive market that drives the decision to call at the specific port. All the shipping companies reported that it would be impossible for them to remain satisfied, or keep their customers satisfied, if calling at a distance from major relevant markets port. This statement holds irrespective of cost and time reductions, productivity gains, and quality performance efforts that might develop by other market players. The presence of "highly skilled labour, administered with flexibility" is also mentioned as a drive that adds considerable value.

As regards the relations with terminal operators, the identified value drivers for shipping lines include the stevedore's experience, confidence in the stevedore's capacities, and the level of adjustments the terminal operator is ready to make in order to suit their needs. When considering the level of adjustments, modernization of technical equipment and extra investments are not always seen as leading *a priori* to a superior outcome. For example, a shipping line may prefer manual, rather than automated, container-handling equipment because "in that way the desired flexibility gains are more probable to be achieved".

When shipping lines do not invest in developing dedicated terminals, the value drivers are more operationally oriented; the provisions of guaranteed space and operation stability are important. Operational stability is most important for the shipping line when its relation with the common user terminal operator goes beyond a typical user-service providers' relation. These are the cases when the port is operated by a sister company or subsidiary, obliging the shipping line to use the particular terminal when calling at the port or to deliver specific volumes per year with no other advantages being present. Stability enables to know in advance the productivity to be achieved and choose the other partners in the chain accordingly. Notably, shipping companies do not necessarily report an experience of vital productivity difference between global terminal operators and terminal operations that their own subsidiaries provide.

In the second setting, for shipping lines that invest in dedicated terminals to be used as hubs and points of consolidating high volumes of containers, the geographical location of a port insofar its capacity to facilitate the offering of global services stands as the most important value driver. Additionally, shipping lines are very interested in the social, economical and political conditions of the country in which they are investing. Although profoundly desired, service quality and productivity are not top priorities. This is because investments are mostly strategic decisions seeking higher levels of independence and autonomy via the creation of internal transportation channels. Eventually, the interest in the roles and services of the other stakeholders is less, as operational interactions with other actors apart from the PA, produce a minor value generation impact. The statement by a representative of such a company is illustrative:

> We only care for the existing socioeconomic conditions within which a port operates and some of the port structural characteristics, such as geographical

> location and connection with the hinterland. The rest is our responsibility; we will take care of them ourselves, and decide which are the best ways to develop our activities.

The port's structural features influence value, as they are directly linked with their investment choices. The state of a port's connectivity with the hinterland and the presence of multimodal routes, i.e. rail, road, inland navigation, are the first mentioned features of importance for that category of shipping lines.

All the shipping companies offering global services note that they remain interested in investments towards the development of terminals even though efforts in the past may not have produced a successful outcome. Such efforts are directed towards the development of internal transportation channels. For shipping lines and any other entity that participates in these channels the most relevant external issues are the port's structural characteristics and the quality of the PAs' activities. As confirmed elsewhere (Carbone and De Martino 2003) the level of integration among the actors involved positively influences these parties satisfaction.

There are cases that, while a shipping company has the scale to develop a high level of autonomy, the volumes of containers that the company transports via a specific port are small for such an investment and effort. Moreover, numerous smaller scale shipping companies do not have the ability to create such settings. In these two cases, shipping lines do not prioritize situations that provide high levels of autonomy. On the contrary, they seek these ports where the other active actors are competitive and achieve high performance level. For these entities operational interactions are extremely important, as their performance is largely dependent on the performance of their own service providers. In line with the connected network theory these companies seek situations that enable them to enjoy value in order to keep their own customers satisfied.

In the case of feeding services, value is "cargo driven" with all value drivers closely related to the availability of cargo and the rates for its transportation. The cost of services provided and the time spent in the port are the next considerations, followed by the guarantees of entrance and services in the port. The pursuing of regular commercial and operational agreements of cooperation, geographical location and notably productivity issues are less significant. A representative of a shipping line of this category, operating in several European ports, stated: "I would reach even the worst port in terms of productivity if rates were good and there was a demand for transportation".

4.2 Freight Forwarders

For freight forwarders the level and quality of direct communications with terminal operators is a major issue inextricably linked with the value they derive from the shipping companies' activities. The absence of congestion, in both the water and land transportation, is by far the most important value driver. Short and reliable waiting times in a terminal are vital. Port proximity with the market they (target to) serve, productivity levels, and low costs also have a significant influence. Major value issues also include smooth customs clearance, multimodal connections with the hinterland, and a substantial number of terminal operators leading to intra-port competition.

Value is also generated via the shipping companies' satisfaction, as the connected network approach would predict. In this case, the key drivers for freight forwarders include (a) the reliability of shipping services; (b) the potential of cargo space commitments; (c) prices; and (d) transit time. The (e) percentage of transhipment cargo is also crucial, as transhipment usually increases transit time. Freight forwarders also seek (f) a balanced combination of the vessel size and the number and order of ports of call in a loop.

4.3 Shippers

The shippers' direct interactions with port service providers diminish, as most shippers use the services of freight forwarders. However, there are industrial companies of a considerable size, such as automobiles and supermarkets that retain control of their own transportation channels, or at least an active role upon organizing and coordinating these channels. These big shippers enact agreements with at least one shipping company for guaranteed capacity in specific routes at given rates. On the land-side of transportation, those major shippers that do not posses their own transportation assets prefer assigning this operation to shipping companies' subsidiaries that specialize in logistics services.

For shippers the key value drivers include: (a) the lead time; (b) accuracy, so that the container will be on time at the place of delivery); (c) the sequence that the port of call has on a given loop, with a desire to use the first port of call in order to minimize the possibility of accumulated delays during the earlier visited ports; (d) the total cost of transportation; and (e) the exact days of arrival and departure. Due to the importance of accuracy, additional costs do not produce dissatisfaction when related to quality services. No matter how satisfied shippers are from a port, they "always" use a range of ports rather than a single one, so that they might downplay the risk of a possible delay. The emphasis is on the overall cost of the entire supply chain, as any negative/positive effect of any port rates may disappear during the inland part of the transportation.

4.4 Port Authorities (PAs)

Port users reach a consensus regarding the PAs' characteristics that act as value drivers. PAs create extra value via activities that attract additional users and cargo volumes. Such activities are developing and leading marketing and promotion that improves the image of the port. When developed in cooperation with existing port users, these activities are conceived in order to produce win-win situations via economies of scale that ultimately increase the relevant port hinterland. An additional value driver is the extent that commercial agreements with port users are flexible, cooperative, and free of unnecessary bureaucracy.

Port users also call for a more active PA role in resolving communication and coordination problems, like those between freight forwarders and terminal operators. The quality and availability of information exchange is regularly quoted as a vital value driver. Representatives of different companies including shipping lines of all sizes, freight forwarders and shippers, emphasize that their own business advantage is in practice transformed and relies on information management business that is coordination, exchange, handling, and diffusion of information. Day-to-day responsibilities on operational aspects are increasingly more automated and standardized, but they have to handle quite a large amount of information in order to provide services that keep their own users satisfied. The following statements are indicative: "Information is becoming the key driver in our community. Ports are driven more and more by that info element and investing in hiring big info handling systems" (CEO of a shipping line). Furthermore, an executive of a freight forwarder company states that "Actually, nowadays we are selling information. We are selling a certain kind of feasibility and our business is more and more reliant on communication and getting the right data".

The coordination of port activities is in close reference to information exchange and part of a strategy that creates communication channels between the various service providers, in turn creating a more efficient use of available resources by the actors involved. The PAs' ability to orchestrate this coordination and maximize the performance of the overall setting is vital. After all, the PAs' position within port networks enables them to be regarded as business ecosystems that facilitate port operations by using critical assets and having both the ability and the incentive to promote the development of value creating activities (van der Lugt *et al.* 2009).

Surprisingly, PAs were found by this study to consider their contribution to information exchange and operation of EDI systems as not being part of their core business or operational activities. The importance of intra-port competition is acknowledged, yet it is claimed that the minimum efficient scale for its introduction in terminals is the threshold of five million TEUs throughput. Another contradictory to port users' perspectives statement from an interviewed PA is that flexibility should not characterize their way of doing business, as "[I]f you are too flexible then you are not a good PA". PAs also reported mechanisms and channels of communications with port users that are well established and accessible by users at any time, like report complaints. Interestingly, none of the other actors recognized the presence of such channels of communications and, moreover, none reported a regular formal communication with the PA of the port in use.

5. Conclusions

The ports' integration in supply chains emphasizes the need to rethink the port (Olivier and Slack 2006) and develop comprehensive port performance measurements that systematically monitor whether ports efficiently and effectively serve their users. The conceptual discussion established the importance of actor-based research in order to study the under-researched effectiveness component of port performance, and revealed that port actors relations fall into a B-2-B context. This study adopted a similar research approach to the literature, which calls for a replacement of the satisfaction construct with a value construct and for emphasis on relational specific evaluations rather than only on transactions.

The empirical findings from four major European container ports made possible the mapping and identification of the complex interactions that are enacted by the involved stakeholders. The interactions that take place correspond to two distinct port settings, with interactions themselves classified as operational or commercial, in a way that assists a future value assessment. Hybrid relations are also present in both settings.

The conducted research identified the key value drivers for the different actors based on the interactions they involve. It also exposed that different port users prioritize different value drivers depending on the setting and the role they assume within the established network. Value is also highly idiosyncratic and varies from one user to another, due to the presence of important differences in relations and requirements, and is also subject to interpretation by the actors themselves. All these are in line with those advocating the limited validity of any generalizations, regarding actors or "one driver that fits all" to be applied in all ports and cargo types (Hall 2004).

At the same time, the research made possible a detailed list of the key value drivers as prioritized by each actor in each setting. There is evidence of a mismatch of port users expectations from the PAs and PAs' priorities, in a way that confirms the claim put forward in the beginning of the chapter: PAs need to work further towards a systematic monitoring of whether their activities serve efficiently and effectively the interests of their users.

At a practical level, this list provides a background for a search towards structures and strategies for capturing opportunities created by meeting the users' unmet demands. For researchers, this list provides considerable proxies that might be used toward the development of a port user's satisfaction tool. Still further case studies are essential in order to enhance the validity and reliability of these findings, i.e. by confirming the types of settings observed, the trades and ports where the identified settings and value drivers are applicable.

In both the "common user" and the "dedicated terminal" settings the various entities involved are interdependent when it comes to achieve and increase the value generated for each of them. As service qualities are highly influenced by a number of interactions, stand-alone initiatives and practices cannot produce maximum outcomes. The coordination of stakeholders via collective initiatives, adaptation mechanisms, mutual understanding, and networking emerge as both appropriate and beneficial (see Chapters 7 and 10). When all these are not limited within the port setting but expand throughout the entire supply chain, they guarantee the successful connectedness of the port, and therefore the generation of value for all users at all stages of the flow of goods to their final destination.

Chapter 9
Improving Port Performance: From Serving Ships to Adding Value in Supply Chains

Anthony Beresford, Su-Han Woo and Stephen Pettit

1. Introduction

Traditional ways of measuring port performance have focused on servicing ships, cargo handling, and equipment and asset utilization. While these measures remain important, they now explain only part of the *raison d'étre* of ports. This chapter examines the relationship between the broad concepts of port-focused logistics services and the specific measures of port and terminal performance. How ports evolved from simple nodes between sea- and land-access into logistics centres in rapidly changing logistics environments is highlighted. The case of Associated British Ports (ABP) Connect, which offers tailored logistics services to both port users and non-port users in the United Kingdom (UK) illustrates this evolutionary path that many ports have followed. Ports now invariably adapt their role to suit the needs of particular supply chains, of which the ports form an integral part.

Traditional ways of assessing port performance often remain valid, but some measures need to be designed and adapted to take account of the current role of ports as logistics centres. A Structural Equation Modeling approach is used to examine whether the logistics services offered by ports can materially improve their overall performance as traditionally measured. This approach was used in order to be able to incorporate a wide range of subjective performance measures, compared to traditional approaches which use objective measures (see Chapter 8).

2. Port Evolution in Changing Logistics Environments

Of all supply chain components, ports face some of the most diverse challenges as they adapt to new commercial environments. These include the introduction of supply chain management, global sourcing, logistics outsourcing and new flexible business practices. To cope with these challenges, ports have adopted a variety of new strategies. Many studies have examined changes in the logistics environment and the responses to these changes by the port industry. Some literature focuses on how ports have coped with the challenges (Notteboom 2004a; Ducruet *et al.* 2009) and have evolved (Beresford *et al.* 2004; Slack and Wang 2002), and some studies suggest how ports should deal with the challenges (Notteboom *et al.* 2009; Notteboom 2002a; Robinson 2002; Heaver *et al.* 2001; Heaver 1995).

2.1 Functional Adaptation of Ports

Port management and strategies are directly and indirectly influenced by prevailing logistics trends since the demand for port services is a double derived demand (Marlow and Paixao 2003). That is to say, demand is for products in the first instance; this derives a demand for transport or shipping which in turn derives a demand for port services. Varying logistics strategies and demand patterns for manufactured goods involve many challenges for the logistics and transport industry which in turn must respond to these challenges. From the port industry's point of view, the shipping industry is that which has the most immediate and direct impact. One of the main strategies which is of particular interest is the integration of ports into the supply chain through activities such as value-added service provision, cooperation with other supply chain members and intermodal service intensification. The shipping industry itself has to respond to change which in turn sets major challenges for the port industry. Specifically, imposed changes such as technological advancements or commercial shocks require port operators and port authorities to take certain appropriate actions (Beresford *et al.* 2004). Especially important are the dynamics among consumers and between consumers and providers of logistics services. In this chapter relevant aspects of port evolution are identified from the existing literature and from ongoing current research (Woo *et al.* 2008). Figure 9.1 summarizes the main elements of marine based logistics chains and the interactions between those elements.

It has been widely demonstrated that manufacturing companies realize the necessity of managing supply chains effectively in response to the globalization of the economy and intensifying competition. As supply chains have lengthened and become of necessity more complex, the management of these chains has emerged as a key part of linking suppliers to consumers. Sophisticated Supply Chain Management

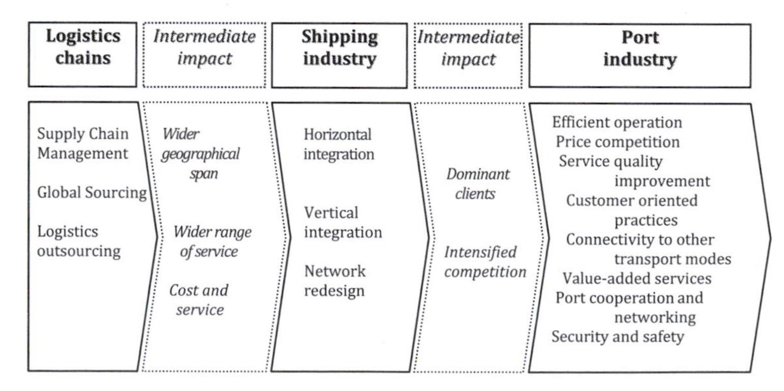

Figure 9.1 Port Evolution in Changing Logistics Environments
Source: Adapted from Woo and Pettit 2009.

(SCM) solutions have been adopted including the outsourcing of certain functions such as logistics (Lambert and Cooper 2000; Rabinovich *et al.* 1999) and transnational ownership of certain parts of the supply chain which are seen as commercially or strategically attractive (Frémont and Lavaud-Letilleul 2009). Such new strategies force transport companies to serve much larger geographical areas and to provide a wider range of services to meet increasingly diversified demand patterns, but at a lower price with higher quality than before (Heaver 2002; Slack *et al.* 1996). To deal with these requirements, in turn, shipping and transport companies have integrated horizontally through mergers, acquisitions and strategic alliances, and vertically through operating dedicated terminals and by providing integrated logistics and intermodal services (Notteboom 2004a). These multiple responses have been driven largely by pursuit of economies of scale and scope, leading to pooling of expertise and capability and to lower unit costs along the whole supply chain.

Additionally, shipping companies have rearranged service networks with the dual aim of global coverage and diversification. The reactions of shipping companies ultimately affected every facet of the maritime industry, especially concerning port operations (Slack *et al.* 1996; Notteboom 2009b). A major transnational study, WORKPORT, carried out in the late 1990s for the European Commission, DG TREN (Beresford *et al.* 2000), assessed working practices and other elements of port activity. This demonstrated that the European port industry, and ports as individual entities, follow different evolutionary paths in response to a series of internal and external influences (Beresford *et al.* 2004). More recently, the concept has been extended to include the role of ports in logistics activities and, specifically, the fit of a port or terminal into a specific supply channel (Pettit and Beresford 2009). The evolutionary dimension of ports' roles in global supply chains is suggested in Table 9.1. Integration clearly does not take place equally in every port and the degree of integration varies by cargo type, trade flow and the wider range of overall commercial requirements.

Table 9.1 The Role of Ports in International Supply Chains

Decade	Ports in Supply Chains
1960s	Low Value Added
	Limited Hinterlands for most Ports
1970s	Closer relationship between ports and users
	Cargo Transformation and Improved Value-Added
	Development of Inland Container Depots (ICD)
1980s	Development of Distriparks from the ICD concept
1990s	Integration of ports with trade and transport chains
	High Value Added
	Emergence of Port Clusters
2000s	Vertical integration of ports with global logistics services
	Lean and Agile Logistics
	Port Centric Logistics

Source: Adapted from Pettit and Beresford 2009.

fic example of the way in which tailor made logistics services have emerged orts can be seen in the UK where the privatization of most ports during the acted as an enabling mechanism for fresh business-oriented approaches, and fresh sources of capital were tapped as the port industry became a more attractive industry in which to invest. Although around 60 ports (mostly very small in peripheral locations) remained under local authority control or ran as Trusts funnelling profits back into port development projects, all the major gateways for deep-sea trade transferred into private ownership. An almost immediate leap forward in productivity and efficiency was observed as new flexible working practices were put in place and fresh sources of capital (from the UK and overseas) flowed in (Beresford *et al.* 2004). The triple objectives of liberalizing outdated working practices, improving UK ports' competitiveness on the international stage and attracting substantial non-government replacement funding were thus all achieved (Pettit 2008).

2.2 Port-centric Logistics, United Kingdom

The largest UK port operator, Associated British Ports, which controls around 25 per cent of UK port capacity, developed a strategy to diversify into logistics services through a subsidiary company known as ABP Connect. This has helped to widen the reach of the port group as a business and in effect, to 'grey' the definition of a port. ABP Connect was established as an ABP subsidiary company in 2001. The objectives of this new company at the outset were to offer tailored third-party logistics capability and a diversified service portfolio to both current port users and to new clients wishing to exploit local non-marine logistics capabilities often involving value-addition activities. Thus ABP was able to consolidate its logistics-related services in a single company and onto specific sites within the ports. They were thus potentially able to maximize the benefits of their activities through aggregation and the exploitation of their strategic coastal locations.

The overall vision of ABP Connect was therefore to simplify logistics by combining the cargo-handling expertise within the company with the strength of its wide geographical coverage. In effect, they were offering a 'one-stop-shop' service for the first time. ABP Connect operates transport depots at, for example, Southampton, Immingham and Barking (London) with a coordinating office in the port of Southampton (Associated British Ports 2009a). At Southampton, the Free Trade Zone offers container groupage, storage, order picking and packaging and sophisticated documentation links with customers and other bodies. ABP Connect links with the EXXTOR car handling terminal at Immingham: a 24 hour, seven-days-a-week facility offering short and long-term vehicle storage and on-site value-addition services in accordance with customer needs (Associated British Ports 2009b). This is a classic model for terminal development where service quality is paramount.

ABP Connect also operates the strategically located inland railfreight terminal at Hams Hall (Birmingham) which handles port-related domestic and Channel

Tunnel traffic and offers high volume capability as well as small consignment just-in-time operations for container and car movements. The terminal is linked to Felixstowe (interestingly, a non-ABP port) by a five-times-a-week rail service. It gained an additional licence in April 2004 to operate full customs-clearance on deep-sea cargo enabling it to operate as a fully fledged inland container depot (ICD). As Felixstowe is owned by Hutchison Ports Holdings (HPH) of Hong Kong, the link to Hams Hall as an ABP facility represents a very interesting example of cross-national cooperation within the supply chain, as ABP is itself partly foreign-owned through the make-up of the Admiral Consortium[1] investment group (Mangan *et al.* 2009). As part of this service diversification process, ABP have engaged in the concept of port-centric logistics by building on the ABP Connect initiative. Several cargoes, such as steel, retail goods, fertilizers and bagged cargo, and some container flows, have been identified, where opportunities for focusing several logistics activities onto the port estate have been exploited. ABP suggest that the growing volumes of imports from overseas have stimulated the development of distribution centres at, or near to, ports which in turn has focused attention on commercially and environmentally friendly logistics founded on sea-mile maximization/land-mile minimization and increased rail transport. Of particular interest is the need for improved visibility along the supply chain such as in the case of retail goods imported via deep-sea shipping. Here, several key partnerships have developed, such as Tesco with Teesport, where the UK's largest port-centric logistics operation has recently been completed, and Asda and Marks and Spencer which are both looking at more environmentally friendly logistics solutions using smaller ports nearer to the inland British markets. Finally DP World, in conjunction with London Gateway, are developing up to 9.5 million square feet of port located distribution facilities to capitalize on the proximity of the site to London (Associated British Ports 2009c). The British model, therefore, is interesting not only from the point of view of supply chain integration *per se*, but also in terms of the internationalization of its ownership.

2.3 Supply Chain Strategies

How exactly port logistics services fit into the overall supply chain will vary from case to case. As ports form a key constituent part of many supply chains simultaneously, and as they play a major role in international distribution, their gateway function is important in itself for value-adding logistics services (Notteboom and Winkelmans 2001). The role of ports within the context of global supply chain strategies has been discussed by several authors (Fisher 1997; Christopher *et al.* 2006; Mangan *et al.* 2007). They have shown that integrated approaches to supply chain management place considerable reliance on port operations in a particular chain because the port acts as a critical node through

1 The Admiral Consortium consists of four partners – Goldman Sachs (USA), Borealis (Canada), GIC (Singapore), and Prudential (UK).

which goods flow. The supply chain strategy chosen will depend on a range of factors which can include 'the predictability of demand for products, the lead time for replenishment of stocks and the logistics philosophy adopted, i.e. lean and agile or 'base and surge' (Pettit and Beresford 2009). Various terminologies have been used to describe these strategies and include, for example, 'leagile' (Naylor *et al.* 1999) or 'dynamic approach' (Gattorna 2006). The lean systems approach focuses on a predictable base demand while agile systems are planned to suit periodic surges, for example, seasonal demand peaks such as pre-Christmas stock-up, and high demand events such as product promotions. Often the response by logistics service providers will be to cater for 'base' demand within the organization but to subcontract 'surge' (demand peaks) capacity to additional third-party companies. This has resulted in two independent but synergous cultures aimed at matching supply and demand as closely as possible: in-house logistics capability and third-party activities.

Ports will therefore play different roles within different supply chains strategies and Pettit and Beresford (2009) discuss how ports fit into these. Table 9.2 details this approach, which builds on earlier work by Christopher *et al.* (2006) and Mangan *et al.* (2007), and develops the conceptualization to include the various types of port logistics facility discussed above. The approach adopted by Mangan *et al.* (2007) is extended by Pettit and Beresford (2009) to break down the various strategies into a typology. They show that, for example, ports providing facilities consistent with short lead-time lean imports and exports would fit an approach combining an A1 import and an A2 export strategy, while a port providing facilities serving long lead-time lean imports and exports would fit with an approach combining a C1 import and a C2 export strategy.

Thus it is important that ports offer services and implement strategies suited to the supply chain, or chains, within which they fit (Mangan *et al.* 2007). However, it may not necessarily be appropriate for a port to simply provide an A1 import/A2 export strategy (Table 9.2). The possibility exists that a port's customer base may be more closely aligned with an A1 import profile but for exports the profile may be D2. The strategy adopted will therefore have to account for the location of a given port, as well as the form of its infrastructure and superstructure. Large ports, many of which act as hubs, such as Rotterdam and Shanghai, have the land area and spectrum of facilities to fit into several strategy combinations simultaneously (for example A1/A2, A1/D2, B1/B2, B1/C2 and so on). Conversely smaller peripheral or spoke ports may only fit into a single strategy type such as D1/D2. The requirements of the shipping companies utilizing the port facilities may also influence the strategy adopted by the port. A modern dedicated container terminal is likely to only provide facilities that service a limited number of strategy types, e.g. A1/A2 and/or B1/B2. Older ports, however, often have surplus land and a wider range of equipment types which mean that they are able to provide services for a greater range of cargoes, thus extending their activities into the C1/C2 and D1/D2 strategy types. Thus, as supply chains evolve, it is likely that ports will become

Table 9.2 Port Roles in Varying Supply Chain Strategies

Supply and Demand Characteristics	Pipeline Strategy	Port Distribution Facility	Ports Role	Strategy Type
Short lead time, predictable demand	Lean, continuous replenishment	Inland Container Depot (ICD), Distripark; ABP Connect	Import: Vendor managed inventory (VMI) at port warehouse	A1
			Export: VMI at the port for short sea traffic	A2
Short lead time, unpredictable demand	Agile, quick response	Distripark	Import: Warehousing, cross docking for rapid import and distribution	B1
			Export: Storage at the port rather than the factory	B2
Long lead time, predictable demand	Lean, planning and execution	Traditional port warehousing	Import: Cost effective storage facilities	C1
			Export: Buffer stocks for seasonal products and varying ship departure times	C2
Long lead time, unpredictable demand	Lean and agile production, logistics postponement	ABP Connect, Inland Container Depot	Import: Warehousing, and manufacturing postponement	D1
			Export: Capability to handle and store non-customized product	D2

Source: Adapted from Christopher *et al.* 2006; Mangan *et al.* 2007; Pettit and Beresford 2009.

more closely bound into them and the concepts that underpin manufacturing and supply chains will become more readily accepted by port and terminal operators.

3. Port Performance Measurement

3.1 Performance Review

Ports clearly play a key role in the operation of supply chains and it is therefore necessary to understand how their performance impacts on the wider supply chain. Beamon (1999) states that performance measurement research often focuses on analyzing performance measurement systems that are already in use, categorizing performance measures, and then studying the measures within a category; much of the research then builds 'rules of thumb' or 'frameworks' by which performance measurement systems can be developed for various types of systems. Performance measures can be reviewed from both 'efficiency' and 'effectiveness' perspectives. Effectiveness can be defined as the extent to which goals are accomplished and efficiency as the ratio of resources utilized against the results derived (Mentzer and Konrad 1991). Efficiency is an objective measure, and effectiveness is a subjective measure. Port performance has largely been measured from efficiency perspectives. Port performance studies can be grouped into two broad categories: a metrics (or index) approach and the frontier approach (see Table 9.3).

The port performance measurement blueprint suggested by UNCTAD (1982) formed a framework for many subsequent studies which drilled down from the

Table 9.3 Traditional Port Performance Measures/Indicators

Literature	Category	Indicators
Metrics and indicator approach		
UNCTAD (1982)	Output	Berth output, Ship output, Gang output
De Monie (1987)	Service	Ship waiting time, Ship's time at berth
	Utilization	Berth occupancy, Berth working time
	Productivity	Cost per tonne of cargo handled
Tongzon and Ganesalingam (1994)	Operational efficiency	Capital and labour productivity, Asset utilization rates
	Customer oriented measures	Direct charges, Ship's waiting time, Inland transport reliability
Talley (1994)	Shadow price	Cargo handling rate, Average delay to ships waiting for berths, average delay to ships whilst alongside berths, truck time working and queuing
Frontier approach		
Roll and Hayuth (1993)	Output	Cargo throughput, Level of service, Users' satisfaction, Ship calls
	Input	Manpower, Capital, Cargo uniformity
Cullinane *et al.* (2002)	Output	Turnover from container terminal service
	Input	Terminal quay length, Terminal area, Number of equipment
Cullinane *et al.* (2006)	Output	Cargo throughput
	Input	Terminal length, Terminal area, Number of quayside Gantry Cranes, Yard Gantries and Straddle Carriers
Wang and Cullinane (2006)	Output	Cargo throughput
	Input	Capital (terminal length), Labour (equipment cost), Land (terminal area)

Source: Adapted from Woo *et al.* 2008.

four measures originally proposed (output, service, utilization and productivity). The 'frontier' approach, on the other hand, measures efficiency in relation to the calculation or estimation of a 'frontier' or 'ideal case', unlike the typical statistical central tendency approach where performance is evaluated relative to an average firm or unit. In the 'frontier' approach, a firm or organization is defined as efficient when it operates on the frontier and inefficient when it operates away from the frontier.

However, several authors (including Bichou 2007; Wang and Cullinane 2006; see also Chapter 7) have argued that port performance studies have traditionally focused on the internal aspects of port operations. This was primarily because the role of ports was seen as merely a node between land transport and sea transport, and the principal aim of ports was to be cost-efficient and provide time-efficient operation, especially in the area of ship servicing; very little if any attention was paid to operations beyond the port gate. Panayides (2002) suggests that ports in the modern supply chain era may have other activities, apart from cargo throughput, which can be measured in terms of performance such as leanness, agility and time compression as well as the performance of other parties in the supply chain. Denktas *et al.* (2009) also suggest, in an intermodal transport context, that when designing performance metrics or when actually measuring port performance:

1. Ports should be recognized as a member of an intermodal transport system and furthermore as logistics centres combining various transport methods, activities, modes and actors;
2. Port operations should be concerned with the process of cargo flow through and beyond the port from entry to exit in either direction from sea to land or *vice versa*;
3. Effectiveness, which concerns the customers' perspective, should be incorporated into performance measurement, and participants in the multimodal transport system other than simply the shipping companies should also be regarded as customers of port.

Added to this list is a further issue: that of transshipment, where the key measure is efficient cargo transfer, ship-to-ship, usually, but not invariably, via temporary storage on land. This common and important activity could generate variants of the key measures suggested above.

3.2 Logistics Activities and Port Performance

The discussion of port evolution and port performance in the previous section raises some interesting questions:

- Do the logistics activities contribute to improvement of port performance?
- If so, do the logistics activities influence both effectiveness and efficiency aspects of performance?
- On which aspects of port operations do the above have most impact?

Several studies provide some clues to answering these questions. Song and Panayides (2008) identified seven parameters for evaluating the extent of the integration and selected variables for port performance. They analyzed the interrelationships between the parameters and the variables using multiple regression techniques. Their results showed that:

1. Information and communication technology positively influence the service quality of ports;
2. The relationship of ports with shipping companies has beneficial, or positive feedback effects on the reliability and responsiveness of ports; and
3. Value-added services are positively related to both port service customization and port service price.

Carbone and De Martino (2003), from their fieldwork interviewing logistics providers and port operators servicing Renault the French car manufacturer, found that Renault outsources some significant parts of the outbound logistics to specialist logistics providers and port operators so as to benefit from the maximum possible service reliability and minimum possible total logistics costs; however, the inbound logistics of component supply is vertically integrated within Renault. This implies that businesses which integrate some logistics functions, e.g. inventory management, with physical transportation, including inland transport and port cargo handling, may produce the highest levels of port performance. Many conceptual and descriptive works also associate the integration of ports with competitiveness or performance issues (e.g. Morash and Clinton 1997; Notteboom and Winkelmans 2001; Paixao and Marlow 2003; De Martino and Morvillo 2008). It is appropriate, therefore, to further examine these possible relationships by means of Structural Equation Modelling (SEM) which is one of the most powerful tools available to scientifically dissect multi-dimensional cause-and-effect relationships and behavioural responses (Gefen *et al.* 2000).

3.3 Research Model and Methodology

The research model was developed based on the discussions on port evolution in Part 2. Eight port evolution aspects were identified from existing published studies on ports, in the form of responses that the port industry has made to the challenges ports face from the structural changes in logistics chains: *Customer orientation; Service Price*; *Service Quality*; *Efficient Operation*; *Safety and Security*; *Connectivity (and intermodal links)*; *Value-added services*; *and Port cooperation and networking* (as shown in Figure 9.1). These were aggregated into three logical groups: external customers' perspectives ('Service'), internal operational perspectives ('Operation'), and logistics perspectives ('Logistics') as shown in Figure 9.2.

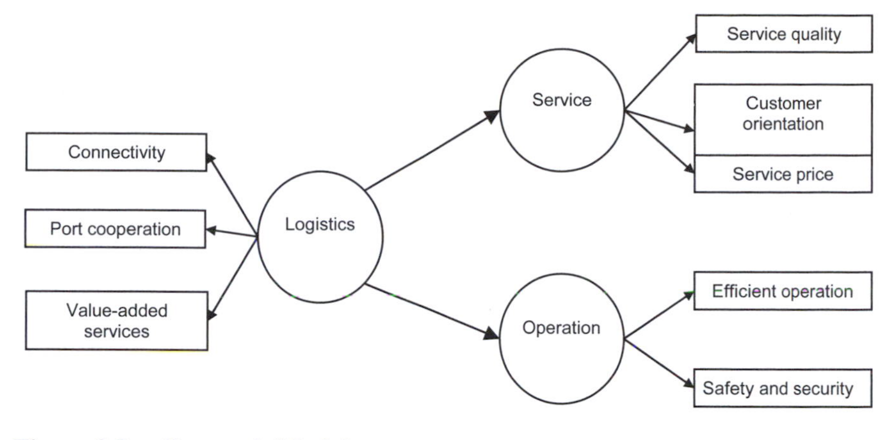

Figure 9.2 Research Model

The 'Service' and 'Operation' constructs represent two dimensions of port performance: effectiveness and efficiency respectively. The 'Logistics' construct represents the logistics activities and supply chain strategies of ports. This research model thus tests two central hypotheses as follows:

- *H1: Logistics activities of ports positively influence the effectiveness of port performance*
- *H2: Logistics activities of ports positively influence the efficiency of port performance*

To examine this research model, two surveys were conducted. The first survey aimed to confirm whether the aspects of port evolution identified in the relevant port literature by port industry commentators and academics are regarded as important. The indicators were then appropriately filtered.

The first survey was conducted over a period of two months from March to April 2008. The questionnaires were sent to 100 potential respondents via email. Before the questionnaire was sent, the questionnaire was pre-tested by two academics and two practitioners in port operating companies. The target population included providers and customers of port services and interest groups in Korea. For the sampling procedure, Korean Port Operating Companies (POC) and Shipping Companies (SC) were selected to represent service providers and customers respectively, and Public Sector Departments (PS) and Academic Institutions (AC) were invited as interest groups. This was in line with the principle that a comprehensive performance measurement system involves incorporating the interests of all relevant members and stakeholders. Of the 100 questionnaires sent out, 72 questionnaires were returned as shown in Table 9.4 (a response rate of 72 per cent).

Table 9.4 Questionnaire Response in the First and Second Survey

Survey	Group	Sent	Not opened	Opened but not responded	Responded	Response rate
First	POC	30	3	7	20	67%
	SC	20	4	3	13	65%
	PS	30	3	1	26	87%
	AC	20	6	1	13	65%
	Total	100	16	12	72	72%
Second	POC	80	12	13	55	70%
	PA	20	12	2	6	30%
	Total	100	24	15	61	61%

In the first survey, respondents were asked to evaluate how significant port evolution aspects are, and subsequently rate the extent to which they should be reflected in port performance measurement from very high (5) to very low (1). The respondents showed very similar patterns of response to both questions. 'Efficient operation' achieved the highest score. This implies efficient operations and efficiency aspects in ports are still regarded as the most important factor in both port management and performance. However, the results also demonstrate that effectiveness aspects such as 'Service price', 'Service quality' and 'Customer orientation practices' are as important as 'Efficient operation'. The lowest score related to 'Port cooperation and networking' which was proposed as one of the important emerging strategies for ports to increase bargaining power in negotiations with powerful clients, and which could be achieved through development of strategic and cooperative networks among terminals in a port or between ports in a country or region. The score was around the medium level (=3) indicating that this aspect should be excluded from the research model. Furthermore, indicators which can stand for, or measure this aspect properly, were unclear in the literature and not widely appreciated by practitioners.

Respondents also rated the appropriateness of the indicators for each aspect of port evolution from very high (5) to very low (1). These were established from the relevant literature. The results show that most of the indicators generated strong agreement amongst respondents about the indicators' appropriateness for measuring aspects of the port evolution. However, indicators whose mean value was lower than 4.0 were excluded in order to simplify the measurement model.

The second survey was designed to measure the port evolution aspects using the measurement instruments developed based on the result of the first survey and was conducted over one week during August 2008. One hundred questionnaires were sent via email and 61 questionnaires were returned (Table 9.4). Managers in Korean port operating companies and port authorities were chosen to represent port service providers. The respondents were required to evaluate the level of logistics activities, services and operations of their terminals or ports using a 5-point Likert scale. In business and operations research, objective measures and data rather than soft measures are preferred for analysis and benchmarking. Researchers have therefore made efforts to search for better concepts and measures for phenomena, activity and performance. However phenomena or aspects of performance which cannot be measured directly clearly exist, including level of integration, cooperation, orientation and collaboration. Also, those which are difficult to measure accurately and objectively include customer satisfaction, responsiveness, and reliability. In addition, researchers often have limited access to some data such as sales and profitability information. Another difficulty is that variations in definitions and data collection procedures and differences in accounting standards make it hard to compare one organization with another. Soft measures may therefore be called for when research faces one or more of the above problems. It is also suggested by Fawcett *et al.* (1996) that even though objective performance measures are preferable, perceptual measures have been found to correlate closely with objective measures. CFA and SEM are appropriate methodologies when valid and reliable soft measures are required.

To analyze the data collected from the second survey, therefore, Structural Equation Modelling (SEM) was employed. SEM is a powerful statistical technique that combines the measurement model (CFA) and the structural model (regression or path analysis) into a simulation statistical test (Garver and Mentzer 1999). The measurement model specifies the relationships between the observed variables and the unobserved variables. Latent variables represent theoretical constructs which cannot be observed directly and observed variables represent measured scores from the instruments. In this study, the port evolution aspects can be latent variables and the indicators for the aspects can be observed variables. The structural model specifies the hypothesized causal relationships among the latent variables. Once the instruments are validated in the measurement model, the structural model is applied to identify causal relationships between the validated latent variables.

4. Empirical Analysis

The collected data were analyzed with the AMOS software package (version 6.0). A sample of this size (61) is categorized as 'small'. Small sample size affects some model fit indices since the fit indices tend to inflate when sample size is small (Hair *et al.* 2010). Missing data were not identified from the collected data. Normality is the final and most fundamental assumption in multivariate analysis. The present data set is generally negatively skewed and AMOS indicates significant skewness. As suggested by Byrne (2001), the bootstrapping technique was employed for remedying the non-normality problems, and this was successfully achieved.

Table 9.5 CFA Result for Service, Operation and Logistics

Construct		Standardized regression weight	t-value	R2	Composite reliability	AVE	Chronbach's alpha
Service							
Service quality	Timeliness1	0.60	4.2***	0.40	0.81	0.52	0.80
	Timeliness2	0.80	5.3***	0.64			
	Lead time	0.70	-	0.50			
	Accuracy of information	0.77	5.0***	0.59			
Customer orientation	Responsiveness1	0.73	6.4***	0.53	0.88	0.65	0.88
	Responsiveness2	0.83	7.7***	0.69			
	Flexibility1	0.86	-	0.74			
	Flexibility2	0.81	7.5***	0.65			
Service price	Total price	0.87	-	0.76	0.90	0.76	0.90
	Handling charge	0.87	8.7***	0.76			
	Auxiliary service charge	0.88	8.6***	0.77			
Service	Service quality	0.77	-	0.60			
	Customer Orientation	0.86	3.6***	0.86			
	Service price	0.72	3.5***	0.72			
Overall Goodness-of-Fit Indices: $\chi2$=63.844(df=41, p=0.013); $\chi2$/df=1.56; CFI=0.94; TLI=0.92; RMR=0.04							
Operation							
Efficient operation	Throughput per crane	0.50	-	0.20	0.79	0.60	0.77
	Ship waiting time	0.99	3.2**	0.99			
	Ship working time	0.79	3.0**	0.62			
Safety Security	Regulation	0.81	-	0.66	0.84	0.72	0.84
	Number of accident	0.89	6.5***	0.80			
Operation	Efficient operation	0.74	-	0.54			
	Safety and security	0.99	2.8**	0.99			
Overall Goodness-of-Fit Indices: $\chi2$=4.695 (df=5, p=0.454); $\chi2$/df=0.94; CFI=1.00; TLI=1.00; RMR=0.03							

Table 9.5 ***Continued***

Construct		Standardized regression weight	t-value	R2	Composite reliability	AVE	Chronbach's alpha
Logistics							
Connectivity	Cargo waiting time	0.85	-	0.73	0.88	0.78	0.88
	Cargo working time	0.92	7.5***	0.84			
Value-added service	Cargo increase	0.82	-	0.68			
	VA increase	0.81	7.2***	0.66	0.89	0.72	0.89
	Adequate facility	0.91	7.9***	0.83			
Logistics	Connectivity	0.99	-	0.99			
	Safety and security	0.79	5.7***	0.62			
Overall Goodness-of-Fit Indices: $\chi2$=10.911 (df=4, p=0.028); $\chi2$/df=2.7; CFI=0.96; TLI=0.91; RMR=0.03							

Note: *** p<0.001; ** p<0.01.

Table 9.6 **Discriminant Analysis**

	Service quality	Customer orientation	Service price	Efficient operation	Safety and security	Connectivity	Value added service
Service quality	0.72 (0.52)						
Customer orientation	0.57**	0.81 (0.65)					
Service price	0.46**	0.54**	0.87 (0.76)				
Efficient operation	0.33**	0.44**	0.46**	0.77 (0.60)			
Safety and security	0.28*	0.57**	0.65**	0.55**	0.85 (0.72)		
Connectivity	0.24	0.49**	0.65**	0.65**	0.56**	0.88 (0.78)	
Value added service	0.37**	0.51**	0.62**	0.55**	0.45**	0.69**	0.85 (0.72)

Note: ** p<0.01; * p<0.05.

4.1 Measurement Model

The overall model fit was assessed using fit indices from various families of fit criteria: chi-square statistics (χ^2) and normed fit chi-square statistics (χ^2/df), Comparative Fit Index (CFI), Tucker-Lewis Index (TLI), and Standardized Root Mean Square Residual (SRMR). The generally recognized criteria of these indices are as follows: $\chi^2/df < 2$, CFI>0.90, TLI>0.90, SRMR<0.08 (Hair *et al.* 2010). The CFA results for 'Service', 'Operation' and 'Logistics' in Table 9.5 shows that the overall model fit for the three measurement models is good.

All items' loadings on their corresponding constructs were acceptably high and significant at the 0.001 or 0.01 significance level except for an item in 'Efficient operation' construct with a loading of 0.50. However this item does not appear to harm overall model fit. This demonstrates adequate convergent validity. The values of Chronbach's alpha (>0.7), Composite Reliability (CR, >0.7), Average Variance Extracted (AVE, >0.5) indicate that construct reliability was confirmed for the measurement models.

Discriminant validity is defined as the extent to which the items representing a latent variable discriminate that construct from other items representing other latent variables (Garver and Mentzer 1999). This offers a method of comparing AVE of each first order construct with squared inter-construct correlations. Discriminant validity is verified as shown in Table 9.6. Firstly, no inter-construct correlation (values off diagonal) higher than 0.85 is found, and the AVE of all the constructs (values in parentheses on diagonal) is higher than 0.5. Secondly, the square rooted AVE of all the constructs (value on diagonal but not in parentheses) is greater than the inter-construct correlations. Therefore discriminant validity was also confirmed among the latent variables examined in the current analysis.

4.2 Structural Model

The structural model was constructed for testing the hypothesized causal relationships. In constructing this model, a partial aggregation method, which uses composites of 2–4 measurement items as observed variables for their corresponding latent variables, was applied to reduce model complexity and identification problems (Leone *et al.* 2001). The individual paths in the structural model were also evaluated as shown in Table 9.7. The path 'Logistics' – 'Service' represents hypothesis H1, proposing a positive causal relationship between the two constructs. H1 was supported with the regression weight of 0.83 (standardized weight of 0.95) and the critical ratio of 6.4 ($p<0.001$). The 'Logistics' – 'Operation' path was also supported with the regression weight of 0.65 (standardized weight of 0.95) and the critical ratio of 6.0 ($p<0.001$). In conclusion, 'Logistics' has a strong and positive impact (0.96) on both the 'Service' and 'Operation' constructs.

Table 9.7 Summary Statistics of Hypotheses Testing Results

Hypotheses	Regression weight	St. weight	t-value	Accept/reject
H1: Logistics →Service	0.83	0.93	6.47	Accepted
H2: Logistics →Operation	0.65	0.95	6.04	Accepted
χ^2=27.03 (df=11); χ^2/df=2.2; GFI=0.91, CFI=0.92; TLI=0.85; SRMR=0.053				

5. Conclusion

There are several implications arising from the integration of port related logistics activities on supply chain performance. The causal relationships between logistics activities of ports and performance were explored and the logistics activities represented by value-added services and intermodal transport services that ports offer. Port performance was evaluated from two perspectives: customer-oriented and internal operational perspectives. The former comprised service quality, service price and customer orientation, and the latter consisted of efficient operation, and safety and security. The research hypothesized that logistics activities contribute positively to both aspects of port performance. The research employed a comprehensive and rigorous methodological process using SEM to assess these hypotheses and the results demonstrate that ports' logistics activities influence both aspects of port performance.

Ports are no longer places where cargo is just stored and they now act as focal points in supply chains where economies of aggregation offer opportunities for value addition or tailoring of services to suit particular customers' needs. Many ports have adapted and developed to allow the supply chains within which they operate to stay competitive. UK ports have not, until recently, focused on value-addition, but rather on cargo handling and ship servicing, whereas continental ports have acted as logistics platforms for more than 20 years. There is now clear evidence of UK ports following the same evolutionary path observed in continental Europe. One clear example of this is that of ABP. The company first diversified by offering value-added services through its Connect division, but has more recently developed further into port-centric logistics by tightening its relationship with major retail companies at specific strategic locations. The development of such sophisticated relationships do not, however, entirely replace traditional port performance measures. As was shown, SEM is an appropriate technique to use in this type of research since it can accommodate the use of both objective and subjective measures.

The use of Structured Equation Modelling caters for both traditional port performance evaluation and modern methods of measuring supply chain effectiveness by using the three parameters of 'service', 'operation' and 'logistics'. The constructs and measurement variables used in this study were successfully validated in the statistical terms although the statistical relationships are not necessarily complete. Future research could expand the number of variables potentially relevant to supply chain effectiveness measurement in order to capture additional dimensions of integration of ports into supply chains. Overall the findings suggest that ports are positively integrating into supply chains in order to hold onto existing business, to acquire competitive advantage and in some cases specifically to diversify their portfolio. Thus, the integration of ports into supply chains takes place to different degrees and in different ways, driven by a variety of motives.

Chapter 10
Coordination in Multi-Actor Logistics Operations: Challenges at the Port Interface

Trevor D. Heaver

1. Introduction

Technological changes in ships and ports and the increased volume of goods transported through ports have increased the importance of coordination at the interface of ports and their hinterland. The increased attention being given to achieving greater coordination is in keeping with long-run and short-run developments. In the long run, shippers have changed their management structures and practices away from a focus on narrow functional activities, such as transportation, to multifunctional responsibilities such as logistics and supply chain management. Also, transportation and logistics companies have extended their reach geographically and across businesses through acquisitions and internal growth. These developments support greater coordination. More immediately, the exceptional global cumulative growth of container traffic from 2001 to 2005 revealed significant inefficiencies and capacity constraints arising, in part, from inadequate coordination and increased by widespread congestion. Changed practices and the 2008-2009 recession have prevented recurrences of substantial congestion but have not removed the need for new initiatives. For example, the US ports of Seattle and Tacoma announced in October 2009 the need to work jointly and with other US West Coast ports and the railways that serve them on new approaches that address the transportation system holistically to respond to the evolving competitive landscape (Port of Seattle 2009). The increased attention given to coordination reflects the importance of relationships among actors to the efficiency of the individual functions, to ports' logistics systems and to the import and export trades of a country.

There have been a number of recent articles and reports that consider issues associated with the level of coordination among organisations and functions in ports. Carbone and De Martino (2003) examine the working of relationship among actors in Le Havre for the traffic of Renault. Song and Panayides (2008) use survey data from port and terminal managers to examine statistically the relationship between measures of integration with perceived levels of port performance. Other studies such as De Langen (2008), De Martina and Morvillo (2008), OECD (2008) and Van der Horst and De Langen (2008) have examined the reasons for sub-optimal coordination among port community actors and approaches to improving coordination. Estache and Trujillo (2009: 69) anticipate a much broader role for port authorities 'as facilitators of inter-modal coordination and of logistics integration'.

The purposes of this chapter are to contribute to this literature and to aid efforts to improve coordination among actors in port communities. The chapter does not offer conclusions but suggests the strategies to improve coordination that may be followed by public landlord port authorities (PAs).

There are three reasons for focusing on the role of PAs. First, as public and independent not-for-profit bodies, PAs enjoy a stature for leadership. Second, PAs are commonly tasked with the role of facilitating trade, as is the case in Vancouver. Third, PAs are now more clearly, than formerly, one organisation in a logistics chain competing against other logistics chains to serve the needs of shippers. Therefore, their interests lie in the efficiency of individual actors and in well coordinated port community functions for the gateway and logistics chain to be efficient.

The effectiveness of coordination among participants in the logistics chain requires a focus on the nature and quality of relationships. This chapter examines aspects of the relationships among the actors in port communities in general but uses as its case study the community of Greater Vancouver where Port Metro Vancouver operates. Port Metro Vancouver (PMV) is the brand name of the Vancouver Fraser Port Corporation, so named in the federal Canada Marine Act. PMV was created, effective January 1, 2008, by the amalgamation of three previously separate port authorities in Greater Vancouver.[1] PMV, located on the southwest coast of British Columbia, is Canada's largest and busiest port. The jurisdiction of PMV extends over nearly 600 kilometres of shoreline in 16 municipalities. The facilities of PMV exist in the fiord-like deep-water Burrard Inlet, the deep-water man-made facilities at the mouth of the Fraser River delta, and the shallow and ocean-vessel accessible facilities on the Fraser River. PMV has 28 major marine cargo terminals, most with on-dock rail facilities, and is served by three Class 1 railways, CN, CP and BNSF. The port is the fourth largest by tonnage in North America. In 2008, PMV reported 82 million tonnes of foreign cargoes through the port of which 68 million were export. The imbalance in this trade is because of the large volume of bulk exports; in millions of tonnes, annual exports were coal, 26; chemicals, metals and minerals, 12; fertilizers, 11; grains, 11; and petroleum products, 7. PMV is also Canada's largest container port, reporting 2.5 million TEUs in 2008. This is over 60 per cent greater than in 2003. The half million autos imported and the 854,500 cruise passengers round out the profile of the port. Although this chapter focuses on the challenges of achieving better coordination among the port's actors, the port's traffic growth attests to the success that has been realised in serving trade. However, this success has increased the pressures to improve coordination.

1 They were the Vancouver Port Authority, the largest of the three, the Fraser River Port Authority and the North Fraser River Port Authority. The record of the corporation of the Vancouver port is the focus of this chapter, but the acronym PMV is used to refer to the organisation in spite of name changes over the years and the amalgamation in 2008.

The term 'coordination' rather than 'integration' is used deliberately in order to avoid the ambiguity that can arise from different interpretations of 'integration'. Integration is commonly used to refer to the reach of ownership of companies horizontally and/or vertically. However, the amount of consolidation in ownership does not automatically equate to the nature and level of coordination among functions. Business units with a common ownership may well be run as quite separate profit centres with very little or no more coordination of functions than might exist between separate corporations.

The chapter is divided into four further sections. The next section reviews the features of ports which pose many challenges to the achievement of high levels of coordination among the various actors. The third section reviews the history of initiatives to improve coordination among the actors in the port community of Greater Vancouver. The fourth section is a synthesis of the tools that may be used by PAs to improve coordination. The chapter concludes with a short note on the role of shippers in coordination in port communities.

2. The Challenges for Coordination

Changes in logistics management generally are marked by the development of new visions and processes, reflected in the use of new terminology. The term 'gateway' was once used mainly by geographers to capture the image of a port serving a hinterland. Now, it has become popular in business and politics to capture the critical role that numerous activities, on and beyond port terminals, play in the flow of goods to and from hinterlands through the port communities. The term 'supply chain management' reflects the heightened linkage and coordination between buyers and sellers and the various organisations, including transport, that are involved in creating value for customers. Logistics systems are vital parts of supply chain structures. Just-in-time and lean-manufacturing systems are approaches to logistics that enable supply chains to function with better coordination among companies and operations using fewer buffer stocks (less inventory) than was required formerly.

While great progress has been made in increasing coordination in supply chains, logistics in and related to ports remains particularly difficult. The need for effective coordination is reflected in greater reference to the 'port community' of all the private and public enterprises contributing to the efficiency of gateways. The challenges for efficient gateway performance are common among ports. They are local problems found globally.

The challenges of coordination among enterprises in port communities have broad conceptual foundations that are reviewed in Van der Horst and De Langen (2008). The authors recognise five general conditions that explain coordination problems. The first condition is the possible unequal distribution of the costs and benefits of investments to improve coordination such that benefits may not adequately compensate some actors for the cost that they incur. The second condition is that the lack of resources may deter or prevent some actors participating, a common situation for small firms. In Vancouver and in many ports, fragmentation is a particular problem in the drayage of containers between port terminals and off-dock facilities. The trucking industry is fragmented, often with many owner operators. The receiving warehouses often have various operating models and ownership relationships. Van der Horst and De Langen (2008) note that these conditions can give rise to the free-rider problem. Third, coordination may be impeded in a highly competitive market by the reluctance of actors to undertake initiatives that would also benefit competitors. Fourth, firms that are risk averse or have a very short-term focus may be unwilling to invest in coordination efforts the results of which are uncertain or are long term. Finally, the lack of a dominant organisation, equivalent to a channel leader in marketing, inhibits the development of strategic initiatives.

There are also many specific conditions affecting the level of coordination in port communities. The conditions vary with the institutional structure of ports and across commodity logistics systems but the following are usually important:

- *The division of responsibility along the logistics chain*. The most important division of responsibility is when the ownership of the goods and responsibility for the logistics system changes, often at a port under terms in sales contracts. Today, this is more likely in bulk than in container systems in which delivered or origin terms of shipment are now common. However, even with retained ownership, changes in functional responsibility occur between shipping lines, terminal operators, rail companies, trucking, warehousing and many other private and public agencies. Coordination problems occur commonly at these interfaces as planning and physical requirements take place within different organisations. Too often, the quality of coordination between the firms in the logistics chain in ports does not receive the attention it deserves from shippers.
- *The mismatch of transport capacities and operating practices by mode*. The capacities of ships, trains and trucks are quite different and the discrepancies are widening as ships, relatively free from route constraints, are getting larger. The inevitable result is increased pressure on the operation of terminals as the interface among the modes. The pressures are exacerbated by the existence of different practices among organisations, especially hours of work.
- *The complexity caused by the number of parallel logistics chains*. Most gateways serve many different types of shippers, sometimes with complementary, sometimes with conflicting interests. The gateways are composed of networks of services that include common and separate elements and those that are complimentary and competing. In Vancouver, the importance of bulk cargoes using the same inland railways as container traffic over routes that face occasional capacity constraints is an important feature.

- *Inadequate information exchange among actors*. Coordinated logistics is dependent on visibility along the chain. Unfortunately, the flow of strategic and operational information among the participants is often inadequate. This may be an historical legacy from times when the importance of coordination was less or it may be because of reluctance to share commercially valuable information.
- *The effects of traffic growth*. Improved logistics has facilitated rapid trade growth, especially in containers. The growth has caused changes in practices and congestion where the expansion of capacity has lagged. For example, port terminals have had to change their practices and utilise their land areas more intensively for container handling which, in turn, has put pressures on the resources of other parties in the logistics chain.
- *The interaction of gateway logistics with the local community*. Increased port traffic, especially the movement of containers by road, has raised noise, air pollution and congestion concerns in cities which, often, are facing increased urban traffic from city growth. The result can be conflicts rather than cooperation with city governments.

Achieving coordination among activities in ports is also made difficult by the number and complexity of the relationships that exist. Actors affect and are affected by relationships along a single chain and by interactions with actors of other chains. Fabbe-Costes *et al.* (2006) note the added complexities in networks where coordination along a single chain must be achieved at the same time as integration is achieved with other chains. The authors characterise the level of integration possible when multiple chains are involved as 'quasi-integration'. They also suggest that the adoption of standards can play an important role in achieving coordination in such networks. Multiple chains are common in ports, including Vancouver, where one of the current strategies is to create greater use of port performance metrics.

3. Response Initiatives in Vancouver

The growth of traffic through ports, the evolution of logistics management concepts and increased global competitiveness have given rise to important needs for greater coordination among actors in logistics chains through ports. Initiatives by actors in parts of chains with relatively few participants occur and can have notable results. For example, the Canadian agri-business company Viterra notes that its success in unloading a record number of grain cars in a day is the result of collaboration with its carrier, Canadian Pacific Railway, especially through improved internal and external communications.[2] However, such examples are recent and, in part,

2 See, Viterra's News Release, October 29, 2008, Viterra's Cascadia Terminal Sets Vancouver 24-hour Unload Record, available at http://www.viterra.ca/portal/wps/portal/viterra.

reflect the influence of many initiatives to advance support for systems approaches to gateway logistics challenges.

The history of initiatives in Vancouver must be considered with recognition of three features of the port. The first feature is the importance of a few bulk commodities. The coal, sulphur and potash trades have had few shippers using unit trains to a few terminals. Coordination was an integral characteristic of these export movements at their inception. Grain exports have a long history in which governmental constraints that are uniquely Canadian inhibited change to a modern well-coordinated system. The problems in this trade have had effects on the efficiency of all rail services. However, with grain aside, the bulk systems through Vancouver have been viewed as efficient and without major coordination issues. Recent concerns have arisen mainly as rail capacity constraints have led to service reliability issues.

The second feature is that, in the container trade, the need for better coordination is a relatively recent imperative. The container business faced some obstacles during its growth, the most crucial of which was the de-stuffing clause which, until 1988, led to containers for local warehouses being trucked to and from the nearby ports of Seattle and Tacoma.[3] Better coordination emerged as a major issue as continued traffic growth has led to congestion in the inland services and as inter-port competition came to be based increasingly on through-transport performance.

The third feature is the organisation of PAs in Canada which inhibited the development of leadership at the local level. From 1936, the port was administered as a part of the National Harbours Board, heavily controlled from Ottawa until the first major relaxation of those ties when the Vancouver Port Corporation assumed authority in 1984. Legislation in 1999 and 2008 devolved greater powers to the local level. However, reluctance of the port authority to exercise leadership was slow to recede. For example, in 1986, when many PAs were providing leadership in EDI initiatives, the new corporation saw no role for itself in a system for which it was not a significant user. No community EDI system evolved.

In spite of the lack of leadership, the Vancouver port community succeeded as businesses, left to their own devices, worked well enough to attract and serve increased trade and to meet inter-port competition. The need for more leadership has evolved slowly and can be accounted for by some broad trends. The gradual change is not surprising; it took logistics management some four decades to evolve into a vital part of corporations' supply chain strategy. First, as corporate supply chain strategies have emerged, there has been a heightened demand by shippers and service providers for services performed more efficiently and with greater reliability than previously. Second, the increase in traffic volumes has resulted in some significant capacity issues and resulted in changed practices and perceptions.

3 The de-stuffing clause required the stuffing and de-stuffing of containers destined for or from local warehouses to be done on the port terminals. This affected much traffic as freight forwarders are important in the Canadian trades.

Third, the example of leadership and collective actions in ports elsewhere, for example, Singapore and Rotterdam, have been important. And finally, as the Vancouver port corporation matured, it has become better able and more willing to play a leadership role.

A brief review of the history of collective actions in Vancouver is an effective way of identifying factors that have contributed to the pursuit of greater coordination and of identifying the roles that may be played by a PA. The original motivation for the collective actions was the common interests of commercial actors, including the PMV, in getting governments, especially the distant federal government, to amend policies that inhibited the Vancouver gateway in competition with the vibrant services through the US Pacific ports. The competition was and is real for some bulk as well as container traffic. It is this competitive environment that has contributed to the early leadership in Vancouver to collective actions.

The first effective voice promoting community interests became the Greater Vancouver Gateway Council (GVGC). This evolved from a coalition, formed in 1987, involving representatives from the port authorities, the railways, employers and port labour to discuss the competitiveness of Vancouver's ports. The discussions led to the formation of the Roundtable on Transportation and the production of a paper on the need for all levels of government to address issues in the taxation of the rail system, primarily fuel and property taxes and capital cost allowances, which resulted in higher railway costs than in the US. The organisation was an effective lobby. Governments were a 'common target' that enabled the different and often competing interests to collaborate. Membership in the organisation expanded and took on a wider mandate following the Asia Pacific Trade and Transportation Forum organised by the Asia Pacific Foundation in 1990. In 1994, the GVGC was formed with the federal Minister of Transport as the Honourary Chair. The GVGC adopted a vision of becoming the Gateway of Choice for North America, intent on achieving that by providing the highest level of customer satisfaction. To achieve the vision, the GVGC sought to identify and address issues that affected Vancouver's competitiveness, and to increase the awareness of the importance of the gateway to the local, provincial and national economies. It has dealt particularly with the need for improved infrastructure.

The success of the GVGC has rested on the need for collaboration for effective action on common problems, initially taxation issues and more recently infrastructure issues. The GVGC has been an effective model in promoting the gateway concept and in funding and executing studies, the most ambitious of which was the 2003 Major Commercial Transportation System Report for the Lower Mainland. This report and related activities have contributed positively to the gateway infrastructure plans adopted recently by governments. It is important to note that the Council does not take up matters that relate to operational matters or business dealings among members, but its work has contributed to better communication among the actors.

The GVGC enabled PMV to advance dialogue within the gateway community without extending its direct influence on businesses beyond its responsibilities.

It became more active in marketing the port through overseas missions with various port actors and it established an office in Beijing. However, prior to 1999, PMV was unwilling to take initiatives that might impinge on private businesses with which it did not have direct dealings. For example, it did not intervene proactively in the dysfunctional relationship between drayage companies and the container terminals.

Subsequently, perceptions changed and PMV became more willing to take initiatives related to gateway activities that may be seen as beyond its responsibilities. An example, dramatic because it was a significant departure from past practices, is its partnership in 2001 with Coast Terminals for the development and management of an off-dock 30-acre container facility for the receiving, unloading, loading of containers and the storage of empties. PMV took this action because it felt that greater off-dock capacity was needed to improve gateway performance, especially reducing the mileage of empty container transport. PMV sold out its interest in the successful business in 2004. PMV truly led.

The self-interest of PMV in fulfilling a leadership role lies in its purpose as laid down in the Canada Marine Act to facilitate trade while remaining financially self-reliant. Improving the efficiency of the gateway is consistent with PMV's mission: 'To lead the efficient and reliable movement of cargo and passengers in a manner that supports Canadian growth and prosperity, now and in the future'.[4] Also, increased traffic is in PMV's self interest as its revenue comes from its leases and from fees on traffic. Nevertheless, it may be argued that initially the organisation was 'pushed' to greater and wider interventions out of necessity.

As in many container ports, the PMV has been in a difficult position with respect to issues surrounding drayage services.[5] The port authority had been an advocate of measures to improve terminal gate performance but it was only following the 1999 work stoppage that it mandated the terminals to introduce a reservations system. PMV managers worked with the stakeholders to introduce systems. Manual at first, a web-based Container Terminal Scheduling System was started in 2001. In 1999, PMV also introduced a truck licensing system by requiring trucking companies to sign a Memorandum of Agreement which raised trip rates and subsequently required a change to hourly rates. However, the rate structure could not be enforced and problems persisted at the terminals and congestion on the roads worsened, leading to another work disruption in 2005. The diversity of interests and the complexity associated with legal jurisdiction led to the settlement through a federal Order-in-Council. However, efforts to prevent a recurrence of the dispute continued. On December 1, 2006, the PMV announced a new trucking policy and a more rigorous licensing, audit and enforcement system.

4 PMV's mission statement available at: http://www.portmetrovancouver.com/about/ourvision.aspx.

5 For insight into some of the issues associated with drayage services see Guiliano and O'Brien (2008) on the implementation of extended gate operations at the ports of Los Angeles and Long Beach.

It is hoped that this licensing system will ensure a more efficient structure without regulating the industry. Improvements in gate technologies continue to be made by the terminals.

PMV has attempted to facilitate greater information exchange among actors in container activities in the expectation that those actors affected by particular issues will work together to improve processes. The Container Stakeholder Group was established after the 1999 work stoppage. The Group has had various subcommittees; currently, they are the Container Terminal Scheduling Committee (CTSC) and the Intermodal Performance Standards Committee.

The latter committee reflects the adoption in 2005 of a supply chain initiative by PMV. Consistent with this initiative, the CTSC has worked to generate metrics that would measure and track the performance of significant attributes of the container logistics system. The metrics are now posted on a so-called 'dashboard' on a website available to those that register (http://www.pacificgatewayportal.com/Pgpsite/Default.aspx [accessed: May 2009]). The metrics serve as a monitor of attributes of container logistics within the port as a whole and are used to identify changes in performance. The metrics provide a factual basis for a review of performance but, in the end, it is up to the business parties to identify and respond to matters needing attention. The metrics do not replace the need for individual actors to maintain their own metrics as the conditions of movements through terminals may well differ.

The introduction of the supply chain strategy coincided with winters of significant disruptions in rail service. In October 2004, even before winter weather caused serious disruption to services, the Canadian Pacific (CP) and Canadian National (CN) railways entered into a co-production agreement to improve the fluidity of rail traffic in the Vancouver area. The capacity issue peaked in Vancouver, as in many ports, in the fall of 2004 when higher than expected peak traffic of shippers was channelled to established routes. The systems, especially the railways serving the West Coast ports of North America, could not cope. The experience led shippers as well as service providers to examine their capacities and practices.[6]

The Canadian winter of 2004 aggravated capacity problems as extreme cold in the Prairies reduced train lengths and there were line closures in the mountains. The Vancouver container terminal company TSI declared force majeure in January 2005, resulting in the diversion of traffic to other ports.[7] Later, it introduced a system of warning levels related to its ability to handle container flows based on rail fluidity. This metric was to alert the shipping lines and the railways of potential problems. The problems faced by TSI and other port terminals were

6 Shippers developed practices to ship earlier to reduce the peak, to produce closer to market for peak-demand products and to reduce reliance on single shipping routes. Subsequent peak seasons passed without major problems.

7 'Vancouver terminal declares force majeure' at http://www.ufsasia.com/newsdetail.asp?id=45.

terminal congestion from the inability to move containers off terminals and the length of time taken by the railways to recover from rail closures. In 2008, the CN and the CP entered into an agreement with the container terminal companies on service recovery rates should winter weather disrupt service. In 2008, the railways maintained a stock of rails cars in Vancouver to buffer their operations against weather disruptions, but severe weather did not occur. The benefits are system-wide although realised first by the terminals. The costs fall initially on the railways.

Although terminals handling bulk commodities were also affected by rail capacity issues, PMV did not initiate a bulk committee until recently. It now has a Bulk Performance Standards Committee with a membership of the railways and the terminals that handle bulk commodities. The bulk logistics systems had not had the same level of coordination problems as the container systems because there are few shippers moving their products by unit trains to specialised terminals. However, congestion on the rail lines and the priority given to intermodal rail services have heightened terminals' and shippers' concerns about the effects on them of the level of rail service and the capacity of the rail network. The committee is currently working to agree on and then release the performance metrics. The challenge for this committee is that the rail service to the bulk terminals is more differentiated by the particular product and terminal conditions than the rail service to container terminals. Thus, providing metrics that protect confidential information and are useful for the different commodity trades is difficult.

The initiatives of PMV and the port enterprise community have been concurrent with supportive government initiatives. The provincial Ministry of Transportation has sought to be better informed on port issues. It established the Lower Mainland Container Logistics Stakeholders' Forum with a mandate to develop and implement tactical solutions and identify strategic long-term solutions that improve the reliability, productivity and efficiency of the land side container logistics system. The Forum has enabled the provincial government to be more informed and involved with other levels on government infrastructure initiatives. The federal government has also played a role in these initiatives (http://www.bctruckingforum.bc.ca/ [accessed: May 2009]).

The Forum members include the PMV, the container terminal companies and representatives of the trucking and warehousing sectors. It has supported studies of container drayage including a study of the application of GPS based on its use on a limited number of trucks providing data for a simulation study (IBI Group 2007).

In October 2006, the federal government launched a new policy direction with the Asia-Pacific Gateway and Corridor Initiative. The next year, this was expanded into the National Policy Framework for Strategic Gateways and Trade Corridors, developed to advance the competitiveness of the Canadian economy in a trading environment of global supply chains. The Framework has guided policies that have enabled the federal government to work with other levels of government and the private sector to make collaborative investments in logistics infrastructure in Vancouver and elsewhere.[8] The projects undertaken in Vancouver were infrastructure investments highlighted previously in the work of the GVGC. The project, announced March 27, 2009 is, perhaps, the most comprehensive involvement of governments and the private sector. It involves the allocation of land by municipalities, investments by the provincial and federal governments, the railways and terminal companies to facilitate a more effective separation of road and rail traffic leading to a more efficient and environmentally sound transport system for the north shore of Burrard Inlet.[9]

The federal programme has included other initiatives. It also funded the 2007 Asia-Pacific Gateway and Corridor Research programmes, the Workshop which was the basis for this book, and the forecasting research and conferences of the Western Transportation Advisory Council (WESTAC). WESTAC is a non-profit association, founded in 1973, dedicated to advancing Western Canada's economy through improvements in transportation. Its membership spans the modes and sectors and includes service providers, labour unions, shippers, policy-makers and regulators. These initiatives are designed to facilitate and encourage change but only indirectly affect coordination among businesses in the gateway.

4. Lessons from the Vancouver Experience

While the Vancouver port community has been successful in accommodating increased traffic, the growth has put pressure on capacities and on the level of coordination among the actors. Examining the response of specific actors and of the Vancouver port community reveals conditions that have helped to change attitudes and improve relationships. It also provides a background by which to identify the functional roles of PAs that affect strategies to advance coordination.

4.1 Conditions Favourable to Advancing Coordination in Ports

Few actors in a logistics chain and significant separability of one chain from another favour good coordination along chains. These were the conditions enjoyed by the unit-train bulk systems during their development through Vancouver. Increased traffic, including in the multiple-actor container trade, brought about new conditions in the form of congestion within the transport system and within the urban community affecting all logistics chains through the port. The new challenges increased the need for greater collaboration and coordination to improve performance.

---

8 This policy promised a new active involvement of the federal government in the transport infrastructure investment in critical gateways and corridors. The real capacity and competitive aspects of the Vancouver Gateway account for leadership here. It contrasts with the surplus capacity in the Great Lakes waterway where particular north – south facilities are congested (see Chapter 12).

9 For details see: http://www.tc.gc.ca/mediaroom/releases/nat/2009/09-h048e.htm.

It is possible to identify three periods with evolving pressures for greater collaboration and coordination within the Vancouver port community. They are: the existence of external threats; the acceptance of and, even, enthusiasm for a gateway vision; and the investment of public capital.

The most important driver of the willingness of members of the port community to work together has been the existence of external threats. The success of the GVGC in collaborative studies and lobbying was due to the shared concern of the PMV and businesses over the policies of the federal government (the distant 'Ottawa') that handicapped the Vancouver gateway. The effects of the policies were made all the more pressing by the competition from US ports, particularly those in the US Pacific Northwest, which enjoyed more favourable public investment and other policies. The actions of the GVGC were addressed to external and common issues, not to relationships among the actors.

The risk of losing traffic to other routes appears to have been a factor leading the railways to accept the costs of improved recovery rates for services to container terminals after winter service disruptions. The result is a more reliable coordination of operations.

While the existence of external threats has been of major importance to collaborative initiatives, prominence of a vision of a well-integrated port community providing gateway functions to serve global supply chains has been important to the acceptance of individual and collective actions. The vision has been an important foundation of the programmes of the port authority and the federal government. It has contributed to the willingness of actors to participate in PMV and government committee structures and, building on these dialogues, to work outside the committees more effectively on operating issues.

Finally, public capital has been a vital factor in advancing coordinated infrastructure developments. Under the Gateway programme, the federal, provincial and municipal governments have made investments coordinated with private investments to bring about important improvements in the road and rail networks to the benefit of local communities and businesses. The investment by PMV in the off-dock container handling facility is unique in Vancouver. The usual investments of PMV have been joint investment agreements with the railways and concession holders for the developments of new terminals without the PMV investing in the businesses.

4.2 Functional Roles of the Port Authority

The activities of the port authority in Vancouver to improve performance in the gateway have evolved as conditions have changed. They have not been undertaken as parts of traditional functional roles of PMV. However, recognition of those roles may assist the development of PA programmes, even though the specific actions of a PA will differ with the business and institutional structure of a port. The measures taken in Vancouver can be placed in four functional categories. They are:

- Promotional: This does not refer to joint marketing but to efforts to persuade others to improve performance, including through increased effective coordination.
- Operational: A PA may take actions through investments or managerial roles that directly affect operations and improve coordination.
- Regulatory and Contractual: These affect others and may deal with the efficiency of individual functions or aim to improve relationships.
- Intelligence: A PA must maintain an intelligence function by which it gathers and analyses information and data on matters in and affecting the port community. It is from this intelligence that a PA develops its knowledge base and builds its vision, policies and strategies. One of the important aspects of a PA's vision is of the PA as a leading member of a port community, in which the actors are committed to providing efficient, reliable and sustainable gateway services. The vision is necessary to get individual corporations to explore ways that greater coordination can benefit all. The PMV's supply chain initiative is a strategy to implement this vision.

The promotional function: At the heart of this function is persuading other actors in the port community to manage their operations so that they as well as the gateway as a whole serve shippers (meet shippers' expectations) more efficiently. Challenges in the pursuit of this role are identification of the specific objectives for actions and selection of the best means to pursue them. Specific objectives could include:

- Increase dialogue among members of the port community. This is the most general approach to improving the working of the port community. In Vancouver, committees and working groups have been used to serve this purpose. However, follow up on issues arising is dependent on the affected parties, as the committees and, usually, PMV do not have the authority to resolve matters in the business relationships of others.
- Add to knowledge and understanding of operations. A better understanding of specific operations can lead to different attitudes and behaviours by the actors. One way to achieve this is also through committees, especially if supported by government. The committee format is useful for the discussion of general matters, for example, the development of metrics, but may be less appropriate as a mechanism leading to studies of specific actors or parts of the logistics system. However, in Vancouver, the availability of government funding has enabled special studies, as in the study of drayage by the IBI Group that might not otherwise have been supported. The conduct of studies, simulations and demonstration projects would all be ways to increase knowledge and foster change. The introduction of performance metrics is an important part of PMV's effort to increase the knowledge of performance levels and, through this, timely identification and response to contributing conditions. This applies for positive and negative changes in performance.

- Improve chain visibility. Arguably, the most important aspect of relationships in port communities is visibility; that is the knowledge of other parties' current and expected operating conditions. The operations of actors upstream and downstream, or working in tandem with a port business, are vital to the performance of the businesses and of the logistics chain. Visibility is commonly thought of as product tracking in a supply chain but forecasts, that may be for various time horizons, are vital components of visibility. The railways claim that the lack of reliability of shippers' long-term and even short-term forecasts of car requirements is an important obstacle to efficient car supply. For Vancouver, the initiative of WESTAC to improve forecasts is a collective way to approach the problem of long-term traffic uncertainty. However, improving short-term visibility, for example, in 6-7 day forecasts vital for operational coordination, is difficult to tackle collectively; it is more dependent on direct relationships between actors.
- Recognise the influence of uncertainty in logistics relationships. Better coordination among actors in a logistics chain is affected by more than the visibility between their operations. It is affected, too, by the nature of the contractual relationship between them, in particular, how the parties deal with uncertainty. In Vancouver, the increased dialogue among actors is contributing to better recognition of the nature and costs of uncertainty and, perhaps, some progress in treatment.
- Change attitudes. The general objective of PMV in its promotional activities has been to change attitudes within the community at large and among the actors in the port. Local government involvement in, support for or acceptance of infrastructure projects has been necessary to achieving coordinated infrastructure projects. Changed attitudes by actors have been promoted by PMV through the concept of accountability by all parties as the cornerstone of improved performance of the gateway.[10] From this acceptance comes the greater willingness to share information, increase visibility and modify relationships.

The operational functions: Only in joint marketing has PMV customarily worked in a coordinated way with members of the port community. PAs have limited powers to control coordination in the logistics chain because they have limited presence in operations. Traditionally, PMV, like other PAs, has focused on its roles in port planning and related infrastructure development and on its responsibilities as harbour master. Investment in port infrastructure has required collaboration with terminal operating companies through concession agreements, with greater involvement for container terminals than for bulk terminals. In some ports, but not in Vancouver, PAs have been active in advancing communications technologies

10 The importance of accountability was featured by the Past President and CEO of the VPA, Capt. Gordon Houston in his Presentation to The Standing Senate Committee on Transport and Communications, Vancouver, 13 March 2007.

that have benefited the port community. However, the need for greater attention to gateway efficiency beyond port terminals has led PAs beyond their traditional roles. In Vancouver, PMV invested in the off-dock CFS. Unlike some other port authorities, for example, Rotterdam, no investments have been made in inland transport, a need met in Canada under the federal gateway initiative.[11] However, PMV is now grappling with the planning document required of the new corporation. Traditionally focused on the uses of its own lands, PMV is considering how to incorporate into its planning its new relationship with local governments and the port community.

The regulatory and contractual functions: The regulatory and contractual functions of PAs rest in their responsibilities as harbour master and through their concession agreements with terminal operators. The concession agreements are long-term contracts by which a PA may affect and control behaviour. In Vancouver, PMV was able to support the introduction of extended gate hours for trucks through features of the concession agreement and introduced regulations on trucks accessing its public terminals, even though the terminals are operated under concession agreements.

Conclusion on the roles of PAs: Landlord PAs have limited powers to change the level of coordination among the many port-related enterprises. Recognition of the promotional, operational and regulatory roles of PAs provides a framework for considering the tools that may be used to increase coordination.

The general importance of these roles has changed over time and varies among jurisdictions. As PAs have given up their terminal operational roles and become more characterised as landlord ports, they have needed to rely more on their promotional activities. However, the need for better coordination with investments in inland functions is leading PAs to re-examine the use of their regulatory and operational powers.

5. Afterword: The Role of Shippers

When examining the performance of the port community, it is easy just to focus on the visible players in the port. However, important players are the shippers who, today, are expected to be aware of and potentially active in all aspects of their supply chains. In Vancouver, bulk shippers have been identified as having a role in the design of the systems of which they are a part. However, not all bulk shippers have paid the attention needed to coordination involving port activities. In the break-bulk and container trades neither shippers nor freight forwarders have had a traffic volume or local presence to exert a significant influence on the

11 In September 2008 the federal government announced investments in five short sea shipping projects and two road projects in Greater Vancouver. 'Port Metro Vancouver pleased with Federal Government's Short Sea Shipping Funding Announcement', Port Metro Vancouver, News Release, 5 September 2008.

development of the system. Too few shippers actively pursue better coordination. However, as greater coordination within the port community becomes vital to logistics performance and competitiveness, it is important for shippers to be more knowledgeable about and active in the pursuit of greater coordination. More attention, including research, on the activities of shippers in the coordination of port logistics activities is warranted.

6. Acknowledgements

I am grateful to the following individuals for discussions with me. However, I am solely responsible for the content of the paper. Curtis Cloutier, Senior Manager Land Operations, Dennis Bickel, Senior Manager Supply Chain and Dale Thulin, Consultant, Port Metro Vancouver; Morley Strachan, TSI Terminal Systems Inc.; Tony Nardi, V.P. Logistics, Neptune Terminals; Marie-Therese Houde, Director, Network Strategies, Canadian National Railway.

PART III
International Case Studies

Chapter 11

Benchmarking the Integration of Corridors in International Value Networks: The Study of African Cases

Jean-François Pelletier and Yann Alix

1. Introduction

In December 2008, the African Development Bank (AfDB) announced that its discussions with the Libyan authority surrounding the African integration process were advancing well.[1] The construction of a highway between the Libyan border and southern Niger, one of the major projects proposed to achieve integration, is believed to hold considerable potential by Muammar Qaddafi, the Leader of the Libyan Jamahiriya. Considering Qaddafi's position on the creation of the *United States of Africa*, corridor development plans suddenly become pragmatic continental integration tools capable of federating Nations. After all, is that not one of the structuring effects of the construction of railroads in North America during the nineteenth century? In the African case, the political outcomes of this can only be speculative, but for stakeholders of economic development in Sub-Saharan Africa (SSA), these plans foster many expectations.

Although SSA has witnessed encouraging growth in the past years, it is feared that the recent downturn in the global economic context might result in a development crisis (African Development Bank 2009). In conjunction with existing infrastructure problems and very high land transportation costs, declining prices for commodities on world markets become another barrier to the integration of SSA economies into global markets.

Academic value network literature stresses that the capability of businesses to successfully respond to customer requirements is now largely based on inter-network relationships (Christopher 1998). This paradigm can be extended to corridor management where stakeholders have to adopt a collaborative approach in order to provide maximum value to entire supply chains. This is reached notably through superior capacity, visibility, reliability, flexibility and efficiency. Otherwise, competing corridors capable of offering better value to customers are selected. For many African corridors, these leading-edge capabilities remain

---

1 Source: http://www.afdb.org/en/news-events/article/afdb-and-libya-strengthen-cooperation-libyan-leader-receives-afdb-president-1531 [accessed: 12 March 2009].

theoretical concepts. The consequence is that many regions remain difficult to access and many economies are considered to be on the margin of international value networks.

By positioning itself as the leading solution in/out of SSA through superior service, can the Libyan gateway/corridor provide the winning conditions for many African resources to finally integrate with international value chains? Even with superior service can the Libyan gateway/corridor compete against existing solutions and, if so, at what costs? Finally, how does the maritime transport component fit into this equation? This chapter addresses these questions by benchmarking African gateways and corridors leading into Chad and Niger by applying a relatively simple methodology to the specifics inherent to SSA.

2. Methodology

Answering the previous questions entails a two step methodology based on (1) a case study analysis of corridors leading into Niger and Chad and on (2) the benchmark of each corridor.

In preparation for the case study analysis, a literature review was conducted to provide a clear understanding of what is meant by integrating transportation and corridors into international value networks and how this can be achieved. This review is followed by a description of the principal trading activities of Niger and Chad based on reports by United Nations agencies and annual reports of the main industrial transnational companies active in the region. This description provides insight on the nature of the value networks within which the respective economies can expect to integrate themselves.

The case study is based on the analysis and benchmarking of corridors having the ports of Cotonou (Benin), Lagos (Nigeria), Port Sudan (Sudan), Douala (Cameroon) and Misurata (Libya) as respective gateways. Information on the availability and state of infrastructure, distances, delays, operational conditions was collected in two phases. First, digital maps obtained through the Digital Chart of the World (DCW) website (http://www.maproom.psu.edu/dcw) were downloaded and used to model the surface transportation system. Because DCW maps date back to the 1992-1993 period, data were validated with reference to recent paper-copy maps. This information was used to calculate precise distances between nodes of each corridor and to apply realistic operational constraints such as average speed, climatic conditions, administrative burdens or daily driving hours on them. Memedovic *et al.* (2008) indicate that one of the major problems in trying to evaluate the capacity of logistic networks in the least developed counties is accessing reliable information. It can also be added that the inherent annual variability of surface transportation conditions in SSA literally eliminates any temptation to undertake large scale on-site assessments as the results would provide evaluations reflecting only a precise temporal period. The costs of undertaking annual assessment can also become prohibitive.

Table 11.1 Sources of Cost Data Used in the Benchmarking Process

Port	Market	Source
Port Sudan	Abéché	Based on average ton-km cost for the Douala corridor. Sudan and Cameroon have the same LPI for domestic logistics costs
Douala	Abéché	Data provided by freight forwarder
Lagos	Abéché	Data provided by freight forwarder
Misurata	Abéché	Data provided by freight forwarder
Lagos	Agadès	Based on average ton-km cost for shipments to N'Djamena and Abéché
Misurata	Agadès	Data provided by freight forwarder
Cotonou	Agadès	Based on $/ton-km from Cotonou to N'Djamena
Douala	N'Djamena	Data provided by freight forwarder
Lagos	N'Djamena	Data provided by freight forwarder
Cotonou	N'Djamena	Data provided by freight forwarder
Misurata	N'Djamena	Data provided by freight forwarder
Cotonou	Niamey	Based on $/ton-km from Cotonou to N'Djamena
Misurata	Niamey	Based on $/ton-km on the Misurata-Agadès corridor

Source: Authors.

In the present case, information was collected from various company-specific websites, individual travelling blogs, web-based travel guides, and through professional publications such as port guides and timely infrastructure reports prepared by the UN Joint Logistics Centre (UNJLC). This approach based on comparing and triangulating multiple sources obviously results in potential data quality problems. To counter this problem, a second phase consisted of validating the results of the modelling process by submitting them to an international logistics service provider (LSP) with extensive operations in SSA and to Libyan organisations that were specifically consulted to provide and validate corridor information and travel feasibility through the Sahara. Surface and marine transportation cost data as well as port tariffs were obtained through international and local LSPs. When cost data were not available, estimations were made. Table 11.1 presents the sources of cost data.

According to Memedovic *et al.* (2008), 'soft data' such as the *Logistics Performance Index* (LPI) developed by Arvis *et al.* (2007) should be complemented by 'hard' indicators. In an attempt to provide a tool able to correctly compare the logistical capabilities of countries, Memedovic *et al.* (2008) have designed the Logistics Capability Index (LOCAI). This index is based on the quality of logistics services, soft infrastructure, trade facilitation, traditional infrastructure adapted to multimodal transportation and a modern infrastructure.

Although the LOCAI has conceptual validity and its implementation could provide a very comprehensive tool to benchmark the corridors, it is believed that adaptations are required in order to align the proposed measurements with the SSA context where typical western-economy benchmarks do not necessarily

Table 11.2 Factors and Their Respective Weights in the Benchmarking Index

Factor	Weight
Distance from gateway to market	10%
Transit time in days	14%
Logistics performance index	11%
Political stability	5%
Safety security issues	5%
Environmental conditions	5%
Gateway to market costs	50%

Source: Authors.

apply. For example, road density in SSA can be a misleading figure if it does not take into account the proportion of paved road, dirt roads and simple tracks. Many precisions also have to be brought to the proposed measurements before they can be applied and other potential factors should be analysed.

The index proposed herein remains much simpler and practical while reflecting major concerns related to the decision making process that goes into choosing shipping options through gateways and related corridors. Also, the methodological approach is supported and validated by logistics service providers operating in the region and on the corridors of study. The selection of indicators and their relative weights can be easily adjusted for diverse requirements and for the analysis of specific value chains. As such, they respond to the specific requirements of changing operational conditions. Table 11.2 provides the details on factors used to evaluate each corridor.

Distance from gateway port to market is preferred to road density because routing options in SSA corridors are limited. When a rail option is privileged such as in the Douala corridor, the distance used includes this multimodal option. Distance necessarily impacts transit times but it is kept as a separate factor because, as a spatial constraint (Debrie 2001), it is considered to play a role in the decision process. Transit time is calculated in days rather than in hours since travelling by night is dangerous for security reasons. The LPI represents an excellent and readily available factor that gives a good global evaluation of logistical capacities on each corridor. Political stability and safety-security issues on each corridor are evaluated in relative terms and on a national basis. The environmental conditions factor is used to reflect climatic and orographic/geomorphological constraints such as seasonal rains or sand dune movements which can ultimately necessitate the closure of a corridor for prolonged periods. Finally, gateway to market cost is considered to be most decisive factor and is appropriately given a weight of half the decisional balance.

3. Value Networks and Integration of Corridors: A Literature Review

During the past 50 years, academics and professionals have contributed to define and elaborate the tools and strategies of what has successively been called *physical distribution*, *logistics* and *supply chain management* (Ballou 2007; Farris 1997; Kent and Flint 1997). As indicated by Christopher (2005), the development of the 'value chain' concept by Michael Porter and the outsourcing of non-core activities such as logistics that followed resulted in the supply chain becoming the value chain. More recent work done by Spekman and Davis (2004) on the extended enterprise, by Huemer (2006) on value configuration analysis, by Walters (2004) on the holonic network structure, by Hearn and Pace (2006) on value creating ecologies, and by Basole and Rouse (2008) on service value networks all tend to confirm that superior value is provided by interorganisational capabilities. Given the contributions of the above-mentioned authors, the term 'network' is hereby used and preferred to the term 'chain', to define the multiple organisations working towards a common objective of delivering value to demand.

Gattorna (2006) and Wheeler and McPhee (2006) as well as Gereffi *et al.* (2005) insist on the necessity to understand the dynamics of the flows underlying the value creation process. To optimise supply chains, it is necessary to understand this process at the network level. In an extensive literature review on value in business, Lindgreen and Wynstra (2005) opine that value is not only related to goods and services but can also be found in buyer-supplier relationships. This distinction is fundamental in the value network concept because the quality of buyer-supplier relationships, as the ones between LSPs and shippers, can have a direct impact on the performance of a network.

Value in networks is dynamic and evolves with changes in buyer preferences, buyer values and in regulations. For example, a secured supply chain is much more valued today than 10 years ago. Social and environmental responsibility programs have also become qualifying, if not winning, criteria in many transactions. Carter and Jennings (2002) add that these types of programs can typically foster better relationships and make networks more competitive.

Value and what we extend to *added-value* and *value-added logistics* is thus considered here to be much more than a product or a service. It also includes intangible elements such as relationships, knowledge and values that enable networks to reach higher levels of performance.

In SSA, it is thus important to consider the specific cultural and social aspects of transactions that are different from western-style commercial relationships that create value. In some circumstances, the capability of a service provider to effectively develop excellent relationships with regional and even local authorities can be a determinant in the performance of a corridor. Value creation in SSA corridors is thus conditioned in many ways by local stakeholders that are usually taken for granted in European and North-American corridors.

The time when transportation was considered a key activity of logistics (Ballou 1992) is now far behind as the supply chain framework elaborated by the Global Supply Chain Forum considers logistics to be a functional silo of individual businesses (Croxton *et al.* 2001). In this vision, collaborative and integrative SCM strategies are thus often centred on manufacturing priorities/processes and LSP (including carriers) become *auxiliary enablers* (Basole and Rouse 2008).

LSPs as well as the gateways and corridors within which they operate are nevertheless obviously integrated in value networks. Yet the *level* of integration of corridors in international value networks and most of all *how* this integration can be reached are not obvious. This integration thus becomes a decisive factor in the competitiveness of SSA economies.

4. The SSA Market Context and Potential Trades

Niger is a vast and arid country situated in the Sahelo-Sudan zone. It is a landlocked country with an estimated population of 14 million, the majority of whom live along a narrow band of arable land on the country's southern border. Totalling 1.19 million km^2, Niger borders seven countries but relies extensively on Benin and Nigeria for its logistical requirements. In 2008 Niger was ranked 174 out of 177 countries on the UN Development Program's (UNDP) Human Development Index (HDI). The per capita gross domestic product was estimated at USD 612 in 2006.

During the past years, Chad has experienced a succession of socio-political crises which have impeded the economic and social development of the country. Since 2006, rising violence in Central-Eastern Chad (Ouaddi) resulting from confrontations between government forces and rebels opposing the Debi regime has significantly destabilised the country. The result of this is a disastrous humanitarian situation, notably in the Sudanese refugee camps and increasing insecurity. In 2008, the country ranked barely above Niger at the 170th place on the UNDP HDI.

The most recent consistent and readily available trade data in tonnage terms for Chad are available through the UNComtrade service and go back to 1995. For Niger, 2005 data are available. Trade levels for Chad were thus estimated from what is observed for Niger. Trade data for Niger were extracted and a query was undertaken in order to isolate offshore traffic. For the sake of simplicity, considering that the two countries of interest have relatively similar economic structures, it is assumed that their respective economies import and export approximately the same amount of cargo per citizen. With this reasoning, Niger's trade was multiplied by the population ratio of Chad/Niger (0.77) to estimate Chad's trade. Table 11.3 presents the results of this estimation. Note that ore exports were removed for Chad because only Niger exports uranium. Also, Chad's oil exports are not considered here because they are shipped via pipeline.

Table 11.3 Estimated Offshore Trade for Niger and Chad (Tons)

Commodity Group	Niger		Chad	
	Export	Import	Export	Import
Aggregates and construction materials	6	9,948	4	7,660
Arms and explosives	-	2	-	2
Articles of various metals	16	8	12	6
Chemicals	68	9,979	52	7,684
Copper/Nickel/Zinc/Aluminum products	-	312	-	241
Fabrics, textiles and clothes	14,122	24,447	10,874	18,824
Fertilizers	-	6,491	-	4,998
Iron and steel products	241	11,695	186	9,005
Manufactured products	1,317	24,789	1,014	19,088
Other	396	3,488	305	2,686
Pulp and paper	8	2,042	6	1,573
Plastics and rubbers	35	3,427	27	2,639
Food and food products	4,107	440,690	3,162	339,332
Products of animal origin	374	-	288	-
Minerals and ores	3,420	42,297	-	32,569
Wood and wood products	-	115	-	89
Printed products	-	964	-	742
Petroleum and gas products	72	60,528	55	46,606
Total	24,181	641,224	18,619	493,743

Source: UNComtrade (extracted 14 October 2008) and author estimates for Chad.

It is clear that the potential traffic generated by the landlocked regions of Niger and Chad remain relatively limited. These countries are amongst the poorest in the world and their populations have very limited access to international goods and services. Thus, much of the trade is regional and does not require large structured trade corridors to reach international markets. There is nevertheless a clear and very significant need for infrastructure and modern trading practices to help these regions develop their economy and efficiently access international and regional markets. In that sense, the improvement of the transportation infrastructure can surely stimulate regional trade integration.

In the case of Niger, short-term cargo prospects come from:

- Project cargo emanating from gold mining and exploration activities in the Liptako region.
- Project cargo emanating from petroleum exploration activities in the Ténéré and Agadem regions.
- Uranium exports from the Arlit-Agadez area as well as project cargo for infrastructure development.
- Sulphur and sulphuric acid imports for uranium production.

- International aid to various populations.
- In the medium term, the eventual development of a pipeline from the Ténéré/Agadem oil fields will also generate project cargo needs. As economic activities are developed, it is expected that this will induce other traffic. It is assumed that international relief aid will continue to represent the essential part of international traffics to and from Niger.

In the case of Chad, the current situation makes any investment or development activity a difficult enterprise. Chad's mining sector is largely underdeveloped but some studies suggest that there is a lot of potential, particularly for gold, uranium, silver and diamonds. In any case, these projects would mean only the carriage of project cargo for short periods. To the extent where the conflict with Sudan is resolved and that this leads to exploration activities, there exists potential traffic for capacity expansion of the petroleum drilling industry. As well, Tamoil Africa's exploration activities in the areas of Irdiss 1, Idriss 2 and Wadjadou 1 located near the border with Libya will generate traffic. International food aid should remain an important portion of international trade in Chad. According to the World Food Program, food aid to Chad between 2003 and 2006 has gone from 16 thousand tons to 70 thousand tons. It is expected that at least these levels of aid will remain in coming years. With the exception of international aid cargo which represents many hundreds of thousands of tons per year, the remaining potential cargoes represent a few thousand tons on an annual basis.

Finally, it should be noted that the European Union and Libya are actively working on capacity building in central African states. These capacity building programs are expected to contribute significantly to the promotion of economic development, notably in Chad, Niger and Sudan. It is hoped that this will reduce the inflow of refugees and illegal immigrants in Libya as well as Europe. Although traffic emanating from these capacity building programs cannot be defined accurately for the moment, the materials funded by Libya and Europe which are required to modernise central African economies in general will possibly circulate through the Libyan corridor.

5. Benchmarking the Corridors

5.1 Operational Considerations

Trade corridors are benchmarked on their operational and cost characteristics enabling four main traffic generating markets – Niamey (Niger), N'Djamena (Chad), Agadès (Niger), and Abéché (Chad) – to be reached. Because capital cities with their economic and administrative roles require large amounts of transportation, corridors to/from Niamey and N'Djamena are benchmarked. Corridors to/from the Agadès region are also benchmarked because it is expected that economic activities induced by the development of uranium mining should generate transportation requirements. Considering that a large amount of potential

Figure 11.1 Central Africa
Source: Authors.

traffic to the regions studied emanates from international aid requirements, the final corridor benchmarked has the Abéché area as focal point.

Each corridor is defined according to the port of origin/destination through which traffic will eventually pass. Five ports are considered to serve as possible gateways: Cotonou, Lagos, Douala, Port Sudan and Misurata. Corridors are benchmarked against their competitors according to the physical characteristics linking the port to the market area. For example, the Misurata corridor is benchmarked against the Port Sudan, Douala and Lagos corridors to reach the Abéché market. After briefly describing each corridor and its particularities in linking a given market, the benchmarking matrix is presented. Figure 11.1 illustrates the important ports and markets in the area of interest. Table 11.4 presents the principal characteristics of each gateway and transportation corridor.

Beyond the description of corridor transportation conditions shown in Table 11.4, each corridor was evaluated upon its operational advantages in reaching a given market. The results of this evaluation are presented in Table 11.5. For distances, the shortest distance to reach a given market is given a score of 5. The score given to other corridors corresponds to the best score (5) reduced by the percentage of the supplementary distance which has to be travelled compared to the shortest route. For example, the Misurata corridor is 20 per cent longer than the Douala corridor to reach the Abéché market. The score of Misurata is thus

Table 11.4 Gateway and Corridor Transportation Characteristics

Gateway	At the Port	Along the Corridor
Cotonou	• Containerships usually wait 17 hours • Administrative and customs procedures to clear cargo can take 10 days	*To Niamey* • Road distance 727 km to Niger border + 300 km to Niamey = 1,027 km • Time to reach border 10 days because of transit delays • Border clearance and travel time in Niger typically 16 days • Total time estimate: approximately 26 days *To Agadès (via Dosso and Birni Nkoni)* • Road distance 960 to Niger border + 726 km = 1,686 km • Total time estimate: 28.5 days (@50 km/hr, 8 hr/day) including customs and other delays
Lagos	• Typically eight days are required to clear customs • 11 days required to reach consignee's premises within the country	*To Agadès* • Total transit time 26.5 days (@50 km/hr, 8 hr/day) including customs and other delays *To N'Djamena* • Road distance: 1,800 km • Total time estimate: 35 days including clearing customs and administrative procedures to enter Chad through Niger and northern Cameroon *To Abéché* • Road distance: 2,350 km • Total time estimate: 37.6 days made up of the Lagos-N'Djamena + time to negotiate N'Djamena-Abeche road (750km @ 35km/hr, 8 hr/day). During the rainy season, June to September this road is reportedly closed most of the time
Doula	• Typically 19 days to clear the port	*To N'Djamena* • Total distance: 1,650 km • Rail to Ngaoundéré (2-4 days + 2 days transshipment to road) • Road haulage between 2 to 10 days • Total time estimated at 42 to 45 days • Practically closed during rainy season (June-September) *To Abéché* • Road distance from N'Djamena: 750 km (@ 35km/hr, 8hr/day). Often closed during rainy season

Table 11.4 *Continued*

Gateway	At the Port	Along the Corridor
Port Sudan	• No precise data available but delays of five to six weeks to clear the port are reported.	*To Abéché* • Total distance: 2,809 km • Rail from Port Sudan to Nyala (2,100 km) in about 85 hours • Road to Abéché (560 km) but transit time indefinable due to situation in the Darfur
Misurata	• Port and customs clearance operations are of five days	*To Abéché* • Road distance: 2,867 km • Most direct route through Aozou strip not accessible due to mines • Average of 50km/day in Libya and 150 km/day in Northern Chad • 1 day required to cross border • Total time: 16 days *To Agadès* • Road distance: 3,500 km • No road to cross the Ténéré desert makes it necessary to go down to Zinder (south-central Niger) before going north-west to Agadès • Average speeds of 50km/hr in Libya; 150km/day from border to N'Guigmi; 36km/hr to Zinder, and; 60km/hr to Agadès • Total time: 19 days

Table 11.5 Operational Benchmarks of Gateway-Corridor Combinations

Port	Market	Distance (10%)		Transit Time (14%)		Overall Logistics (11%)		Political (5%)	Safety/Security (5%)	Environmental (5%)	Total (50%)
		Value (km)	Score	Value (Days)	Score	Value (LPI)	Score	Score	Score	Score	Score
Port Sudan	Abéché	2,809	4.1	Unknown	0.0	2.71	4.75	1	1	3	23.7
Douala	Abéché	2,392	5.0	47.6	0.0	2.49	4.37	2	3	2	26.6
Lagos	Abéché	2,542	4.7	37.6	0.0	2.40	4.21	1	1	2	22.6
Misurata	Abéché	2,867	4.0	16.4	5.0	2.85	5.00	4	4	3	44.0

Table 11.5 *Continued*

Port	Market	Distance (10%)		Transit Time (14%)		Overall Logistics (11%)		Political (5%)	Safety/Security (5%)	Environmental (5%)	Total (50%)
		Value (km)	Score	Value (Days)	Score	Value (LPI)	Score	Score	Score	Score	Score
Lagos	Agadès	1,649	5.0	26.5	3.1	2.40	4.21	1	1	2	31.9
Misurata	Agadès	3,503	0.0	19.2	5.0	2.85	5.00	4	4	3	36.0
Cotonou	Agadès	1,716	4.8	28.5	2.6	2.45	4.30	3	3	3	35.2
Douala	N'Djamena	1,650	5.0	45.0	0.0	2.49	4.37	2	3	2	26.6
Lagos	N'Djamena	1,800	4.5	35.0	0.4	2.40	4.21	1	1	2	23.5
Cotonou	N'Djamena	2,831	1.4	35.0	0.4	2.45	4.30	3	3	3	22.4
Misurata	N'Djamena	3,608	0.0	19.1	5.0	2.85	5.00	4	4	3	36.0
Cotonou	Niamey	1,025	5.0	26.0	3.5	2.45	4.30	3	3	3	38.1
Misurata	Niamey	3,940	0.0	19.9	5.0	2.85	5.00	4	4	3	36.0

Source: Authors.

equal to 80 per cent of 5. The same type of calculations was applied to determine the score of delays. The final operational score is based on the overall logistics capability of the corridor as given by the World Bank's LPI. Unfortunately, the Libyan Arab Jamahiriya is not represented in the index due to a lack of data. The score given to the Misurata corridors thus corresponds to the average LPI of upper middle income countries such as Argentina, Costa Rica, Poland and Venezuela which have a similar per capita GDP (at purchasing power parity). Specifically, this score is 2.85. To adopt a uniform evaluation process for each factor, Misurata's LPI obtains the best score (5) and other corridors obtain a score based on their comparative result.

As seen in Table 11.5, the Misurata corridor ranks first at the operational level to reach the markets of Abéché and N'Djamena. This advantage occurs because delays are given a higher weight in the evaluation than the two other criteria. It is also the result of very long delays usually witnessed in the ports of competing corridors. As such, the competitiveness of the Misurata corridor depends on its ability to rapidly clear cargoes bound for landlocked regions. Ideally, administrative and customs clearance procedures for products bound to the south have to reach international performance levels, taken as no more than five days of customs and administrative procedures before cargoes can continue towards their final destination.

The Misurata corridor is visibly disadvantaged in terms of surface transportation distances. It should nevertheless be noticed that the performance of continental corridors depends also on offshore supply networks. The fact that Misurata is closer to North America and Europe can substantially reduce the total transit time and delays to reach the markets even if surface distances are more important.[2] In the same manner, the Misurata corridor is disadvantaged from the Port Sudan corridors for all cargoes bound to or entering the markets from Asian origins or destinations.

The Port Sudan corridors are considered to have very low security and political stability levels. At the environmental level, climatic conditions are considered to be equivalent to other corridors and were given a median score.

In the case of Douala, political and security levels are considered to be relatively stable but seasonal rains make the corridor inoperable during four months, thus reducing considerably the efficiency of it. For the Lagos corridor, the political and security situation in the country is very poor but environmental factors are considered normal.

The Cotonou corridor obtains a median score for all criteria and finally, the Misurata corridor obtains the best scores in terms of political and security factors but difficult climatic conditions in the south result in an average score for environmental factors.

---

2 For example, at an average speed of 15 knots, Misurata is more than 2.6 days closer to New York than Douala. The advantage from Le Havre is almost six days. When taking into account the total distance including the marine leg of the trip, Misurata certainly presents a clear advantage also in distance terms.

If the AfDB/Libyan infrastructure development projects into central Africa go forward, the Misurata corridor should significantly ameliorate its position. These projects will make the Misurata corridor even faster and more reliable and for some markets, the distance should also decrease. For example, the proposed road into Agadès from the Libyan border is expected to reduce the total distance from Misurata to Agadès by approximately 1,000 km.

Because of the inherent variability of operational elements, notably for transit times, corridors obtaining scores within a range of about 5 per cent to 6 per cent should be considered as being equivalent. For example, for the Agadès market, the Lagos, Misurata and Cotonou corridors can be considered as being equal in operational terms.

The operational benchmark matrix also hides the reliability factor of trade corridors. Although shippers consider shorter transit times to be attractive, the reliability of this transit time can be much more decisive in the decision to ship products through a corridor rather than through another. According to the information gathered for this study, it appears that the reliability of competing corridors is particularly deficient.

5.2 Economic Considerations

The next step of the benchmarking process consists in introducing the cost factor. For each of the 13 port/market combinations, a cost by delivered container was established. Based on cost data obtained through freight forwarders, the costs of using the Misurata corridor are consistently higher than using other options. In the best of cases, the $/ton figure for Misurata is 126 per cent higher than the best delivery price for a given port/market combination. This seriously handicaps the Misurata corridor, notably for humanitarian food-aid which often has a lower price per ton than the transportation costs through Misurata. There is presently no competition between surface carriers in the Misurata corridor. In that sense, without a competitive environment, operators will tend to charge high prices because they are not pressured by competitors to obtain business.

When considering maritime freight rates in the cost benchmark, results remain unfavourable for the Misurata corridor. First-hand data collected reveal that February 2008 maritime freight rates between Antwerp and Misurata are approximately USD 300 lower per container than to Cotonou. On a cross-Mediterranean basis (origin Italy), the Misurata routing is USD 700 lower per container than for Douala. However, this reduction does not significantly reduce the overall, or total, logistics costs of shipping cargo into the prospected markets. For example, when taking into account both sea and land freight costs, the Misurata corridor remains twice as high as the Douala corridor to reach Abéché for cargo originating from Europe and loaded in Italy. Adding the marine transportation costs from North America, Europe and Asia would thus not change much the conclusions.

The surface transportation costs into Niger, Chad and Darfur through Libya are prohibitive for low value items such as international food-aid which represents the bulk of traffics. Cargo which can eventually support these cost levels are mainly composed of project cargo and uranium but if current supply chains are judged sufficiently secure and not time-sensitive by the shippers, existing supply networks may be maintained even if the reliability is not at its best. Nevertheless, given the high value of project cargo associated with petroleum exploration and mine prospecting activities, these companies are more sensitive to frequency, reliability and operational quality and should be prepared to pay more, especially if the delivery is time sensitive such as medical supplies, drilling heads or even engineers.

When giving operational and cost factors a 50/50 distribution in the benchmarking matrix, the Misurata corridor becomes much less attractive. As shown in Table 11.6, the results indicate that traditional corridors used for markets remain the benchmark reference for transportation requirements. For example, Douala ranks first for the Abéché and N'Djamena markets while Cotonou is the best performing corridor to reach Niamey and Agadès. When putting these results in relation to the potential traffic identified above, it is clear that even in the most optimistic scenario, the Libyan corridor will not be capable of attracting significant traffic levels.

In sum, although the trade corridor through Misurata appears to be an excellent option in operational terms, it remains uncompetitive in its actual state. One of the main reasons that can explain the high levels of surface transportation costs comes from the lack of experience. Because none of the corridors have

Table 11.6 Total Benchmark Results

Port	Market	Operational weight* (50%)	Score	Cost Weight (50%)	Total (100%)
Port Sudan	Abéché	23.7%	4.6	46.2%	70%
Douala	Abéché	26.6%	5.0	50.0%	77%
Lagos	Abéché	22.6%	2.7	27.2%	50%
Misurata	Abéché	44.0%	0.0	0.0%	44%
Lagos	Agadès	31.9%	3.7	37.2%	69%
Misurata	Agadès	36.0%	0.0	0.0%	36%
Cotonou	Agadès	35.2%	5.0	50.0%	85%
Douala	N'Djamena	26.6%	5.0	50.0%	77%
Lagos	N'Djamena	23.5%	1.7	16.7%	40%
Cotonou	N'Djamena	22.4%	3.6	36.0%	58%
Misurata	N'Djamena	36.0%	0.0	0.0%	36%
Cotonou	Niamey	38.1%	5.0	50.0%	88%
Misurata	Niamey	36.0%	0.0	0.0%	36%

Note: * From Table 11.5.
Source: Authors.

been regularly used by surface transportation providers, it appears that numerous uncertainties remain. To protect themselves against these uncertainties, carriers prefer charging very high tariffs. If the corridors into Niger and Chad become regularly used and that uncertainties and risks are minimised, it is expected that surface transportation costs will fall. The risks and uncertainties are essentially associated with transit in Niger and Chad. For Libyan carriers, entering these zones mean, amongst others:

- War risks/attacks by rebels.
- Risks and uncertainties associated with climatic conditions such as rains and sand storms which can damage trails/roads.
- Risks and uncertainties associated with distance to maintenance services.
- The teams accompanying the convoys must be much more experienced and versatile. For example, convoys should include mechanics, technicians and security personnel.

The only way to mitigate these risks and uncertainties is to make tests in the corridor in order to better apprehend them and to find pragmatic solutions.

Another factor which plays against the Misurata corridor is the fact that the calculation of costs was done on a 'one container basis'. In that sense, escort costs are applied to only one container and volume reductions are not integrated in the calculus. Applying potential volume reductions proposed by the various carriers consulted would enable to reduce the $/container figure by about 6.3 per cent but this would still be much higher than alternate corridor options.

Calculations were also made to determine the Misurata corridor costs levels required to reach total benchmark averages similar or close to the ones obtained by the best ranking corridors. Given the uncertainties related to the benchmarking process, an average of 5 percentage points under the best score is believed to be 'in the ball park' to be competitive and capture clients. In the case of the Abéché market, the Misurata corridor would have to be capable of offering transit tariffs of about 38 per cent less than the quotations obtained. For the Agadès, N'Djamena and Niamey markets, tariffs would have to be reduced by 71 per cent, 62 per cent and 86 per cent respectively in order to become attractive for shippers in general.

At this point, the result of shipper/carrier negotiations becomes the cornerstone of the competitiveness of the Misurata corridor. In these negotiations, the high cost of transportation services into Chad and Niger can certainly be explained by the difficult climate, terrain and security situations but whether or not this justifies a 100 per cent cost premium compared to the Douala corridor (for example) remains unknown. It is more than probable that a shipper interested in shorter transit times and higher reliability and having constant and significant volumes would become a very interested party in Libyan carriers. The perspective of new and potentially profitable business should do the rest in the negotiation process.

While reliability and on-time delivery can be viewed as basic elements of service in many parts of the world, they are far from being common on SSA corridors. In this region, the comparative value-added logistics services can be considered as simply as being:

- On-time delivery/respect of schedules/guaranteed delays.
- Frequent departures (once a day, three times weekly or other).
- Secured warehousing.
- Bonded transit.
- Secured transit.

Although the competitive factors affecting the performance of corridors in reaching central African markets are quite different from other regions, lessons can nonetheless be learned from these cases. First, the analysis of market dynamics in Europe reveal that performing trade corridors are the ones within which a high level of collaboration and cooperation between channel partners is practised. Precisely, this means that shipping lines, port operators, freight forwarders, logistic zones and surface transportation providers work hand in hand in order to implement optimal network solutions for their clients. By working as a whole, synergies can be reached and the resources available in the corridor are optimised.

Nothing seems to indicate that the situation will dramatically change in the near future regarding the competitiveness of corridors leading into central Africa. Nevertheless, the possibility that international investments in physical and informational infrastructure as well as capacity development in logistics and customs will impact on the effectiveness of these corridors should not be underestimated. For example, many projects will certainly generate results in the medium term and will necessarily ameliorate the competitiveness of the corridors. For example, the CEMAC – Transport-Transit Facilitation, Nigeria Federal Roads Development Project, the National Emergency Transport Rehabilitation Project (Sudan) and the Emergency Transport and Infrastructure Development Project (Sudan) total close to USD 735 million of investments in the next years.

Also, with the reintegration of the Libyan economy in world trade, one cannot underestimate the role that this country will play in the future. Accompanied by the development of exploration activities in southern Libya and by investments made by Libyan interests in Niger and Chad, the 'northern route' could become more attractive as traffics converge and grow.

6. Conclusion: And What About Integrating International Value Networks?

The integration of SSA corridors into international value networks is clearly one of the major challenges facing these economies. Today, although some sectors have managed to do so (Kaplinsky and Morris 2008), the *hierarchal Global Value Chain* (GVC) governance type appears to yield the best results as control and coordination of the total network is maximised. In their typology of GVC governance, Gereffi *et al.* (2005) explain that the complexity of information and knowledge transfer, the extent of codification of this information and knowledge and the capabilities of actual and potential suppliers define the five global value chain governance types. When supplier capacity and the ability to codify transactions are low, a hierarchal governance structure will emerge. This would also explain why groups such as Daewoo logistics literally aim to extend their control to the resources and the lands on which these resources grow in order to secure supplies.[3] As pointed out by Panayides (2002), transaction-costs economics also indicate that networks characterised by complexity often become more efficient when vertically integrated.

If we come back to our assumption that the term 'network' defines the multiple organisations working towards a common objective of delivering value to demand, this chapter indicates that African corridors do not necessarily share a common objective. Corridor members whether they are port authorities, customs agencies, international LSP or local carriers all have distinct preoccupations that range from the local police officer trying to insure its daily needs by 'taxing' cargo passing through the village to port authorities trying to secure their domain. Although land operations in the corridors are difficult due to social and environmental constraints, gateway delays remain the principal barrier to the acceleration of cargo passage. Cargo clearance procedures at some gateways range easily from 10 to 30 days without any level of reliability on the actual time it will take to move cargo on to the roads. Even if surface transportation is burdensome, at least one knows that cargo is advancing when it is on the road.

At the value level, the lack of coordination implies that organisations wishing to source products in SSA have to evaluate corridor options not on traditional criteria used to benchmark advanced economy corridors but on elements such as reliability, security and uncertainty management capacities of LSP operating in these regions.

Concerning integration, whereas relationships between manufacturing and distribution elements of networks are considered to be at the forefront of superior value delivery by most supply chain management scholars, the purely logistics and transportation elements of SSA networks are at the core of competitiveness. In other words, while LSP are sometimes considered as auxiliary enablers in advanced economies, they become core value creators and strategic channel members in SSA. The growing interest of global carriers, global terminal operators could thus encourage the further integration in international value chains (Alix and Germain 2008). Although they are currently not the major and historic actors of SSA corridors, Asian and Mid-East players appear to be decisive elements in the capacity for these economies to integrate global value networks. By playing a significant role in the recent development of SSA international trade and infrastructure, these stakeholders could become the vector of integration. Nevertheless, it is believed

3 Source: Financial Times in: http://www.courrierinternational.com/article.asp?obj_id=91867 [accessed: 19 December 2008]. This rumour was eventually contradicted by the Malagasy authorities.

that the players capable of structuring stable and cooperative relationships with national authorities, surface carriers and most of all local populations/authorities can yield the highest potential for integration. As explained by Rizet (1997) through a banana export example, large global shippers also hold the potential to ameliorate the integration/performance of SSA in international value chains.

While the World Bank indicates that 15 per cent to 20 per cent of import costs to landlocked countries are associated with transportation costs compared to about 5 per cent in developed economies, the integration of African corridors into international value networks necessarily implies a revolution in the governance of integrated intermodal services in SSA.

The current chapter has demonstrated that a Libyan corridor through Misurata to major Niger and Chad destinations, although potentially valuable and competitive, is not now viable. Much needs to change, not just in transportation infrastructure but in the governance and operation of integrated intermodal services.

Chapter 12
Building Value into Transport Chains: The Challenges of Multi-Goal Policies

Emmanuel Guy and Frédéric Lapointe

1. Introduction

Gateways and trade corridors in Canada and elsewhere are often associated with initiatives to update the networks that support trade. They are planned to ensure that goods can be moved and traded with fluidity. The need for such planning comes about because of fears that freight movements might be impaired. Thus, debates on gateways and trade corridors in the public arena arise because of concerns about the adequacy of transport infrastructures, and the pressures put on them by current – expected or desired – growth rates of freight movements. But concepts like 'corridors' that invoke the passage of goods through numerous components of a system shed light on the importance of coordination. We argue that the fundamental question raised by the Canadian gateway and trade corridor initiative is not the amount or type of infrastructure needed, but how to bring together the numerous stakeholders of the transport chain to better collaborate. This collaboration is essential to ensure that limited resources are used in the most efficient way and enhance the performance of trade networks. This is evident in the way industry stakeholders have maintained their positions despite the present economic crisis marked by falling trade levels that have offset the imminence of under-capacity.

This chapter has two objectives: (1) to consider the gateway and corridor initiatives in Canada as a policy process characterised by the goal to integrate the interests of a vast array of stakeholders; and (2) to consider the gateway and corridor initiatives as one element of the overall public framework for the Canadian maritime sector. As a case study, we focus on the St. Lawrence and Great Lakes basin (see Figure 14.1).

2. Policy Considerations

Policies and the processes by which they come to be created, implemented or updated represent a vast research topic. In this chapter, involving issues of coordination in transportation planning and a consideration of the role of stakeholders' interactions, we work from the following observations:

- Specific policies that have been implemented and the overall public framework they form are the result of successive rounds of a power play among stakeholders. The consensus reached through consultation-proposition-formulation cycles of the policy making process are, by nature, compromises. They are mixes of the solutions and considerations of the different stakeholders that are sought in order to achieve the minimum cohesion required so as to avoid confrontation or refusal from any of the participants, or alliance of participants, with sufficient influence to block the implementation. It is the nature of public authorities to look after different interests creating policies that therefore contain multiple objectives. The policy development process needs to build coherence within numerous objectives as it seeks to move towards implementation. Yet it will normally retain some degree of contradictory objectives, or else it is likely to be deemed one-sided, favouring one interest-group, and risks to be blocked by stakeholders who see their interests insufficiently considered. As a result, specific policy measures are unlikely to be economic optimums, nor the best practical means to achieve strategic goals in a given socio-economic environment (Everett 2005; Guy and Urli 2009; Guy and Lapointe 2008).

This critical characterisation of policy measures and the process through which they are created has a number of implications for this analysis of the gateways and trade corridors program:

- A direct implication of the point above is the respective positions of stakeholders and how they are voiced. Public stands signal any change or evolution from previous positions and how the respective stakeholders can complement or oppose each other. Public stands also inform of the nature and state of issue-based alliances formed among them. These are indications of premier importance in understanding the policy outcomes and evaluating the potential progress they may support.
- New policy measures may take the form of a revocation or transformation of existing dispositions when the stakeholders identify them as problematic. Often though, new policy measures are put in place as responses to new considerations that emerge within society as a whole or a part. Even when new policy measures are closely associated with existing positions, it is likely that the power play amongst stakeholders will focus on specific issues of desired or feared effects of proposed measures. There will be little consideration of their implications on the coherence of the general framework. Over time, the general policy framework for a sector becomes composed of an historical accumulation of specific measures that are unlikely to contain apparent coherence (Guy and Urli 2009; Guy and Lapointe 2008).
- An analysis of varied policy measures aimed at supporting shipping in different socio-economic and political settings suggests that less than

expected performance and dissatisfied stakeholders can generally be traced back to an unclear arbitration of objectives (Guy and Urli 2009). Stakeholders and governing authorities especially are generally prompt to state clear – usually ambitious – goals for their policies. Lack of clarity in objectives comes from pursing competing or not completely coherent goals. This is often found when objectives for a given sector of activity clash with those of the overall economic policy. For example, a large number of economic policies are aimed at supporting the development of particular sectors of activities or specific regions by providing some sort of assistance that can then confer an advantage over competitors. Yet these policies are usually set within larger ideological and legal systems oriented to the protection of a free market. Ambiguities can also be the product of the historical accumulation of independent measures discussed above.

- Policy making is a decision process. However, when using input from decision sciences it must be kept in mind that in a public multi-agent context, decisions are not 'taken' but result from complex incremental interactions among stakeholders (Bots and Lootmas 2000). Classical tools of decision support may be ill-adapted, and their focus on common interests or consensus can sometimes mask diverging interests among stakeholders that are bound to reappear later (Guy and Urli 2007, 2008).
- Policies are first and foremost the expression of stakeholders' aspirations. They define what should be. In the early 1990s, French contributions in political sciences introduced the concept of *analyse cognitive* that suggests that policies feed on mental representations through which stakeholders can define their own actions and legitimacy. Although *a priori* far from the immediate interests of applied maritime research, this theoretical approach formalised by Muller (2005, 2009)[1] casts an interesting light on the power play often observed in maritime policy making. It suggests that the function of public policy is to ensure sector-based representations retain a minimal cohesion with the commonly shared representations of a society, state or supra-state organisation. 'Public action' – a change in policy or requests for change – takes place when a rupture is perceived and an adjustment of the global and sector-based paradigm (*référentiels*) are needed. Therefore, this realignment, or the tensions implied, should be observable in the discourse of stakeholders, giving rise to changing policy initiatives or contemplated options with fundamental implications for the transportation sector.

3. Public Framework for Maritime Transportation in Canada

The conduct of shipping and maritime operations is highly impacted by an array of interlocking legal dispositions, governmental administrative programs and

1 For a text in English see Surel, 2000.

industry-led common initiatives. We refer to this as the public framework for the maritime sector. In this chapter we limit our analysis to the Canadian framework. But this national system is obviously very much influenced by international realities, starting with the conventions signed under the auspices of the International Maritime Organisation. To analyse this public framework it is useful to divide it around triptychs made of policy goals, effective measures, and impacts that we derive from our considerations of the nature of the policy process discussed in the previous section. Hence, we regroup policy elements in three main families according to their objectives: protect peoples and the environment; create wealth within the Canadian maritime industry; and create wealth in Canadian society through international trade (Guy and Lapointe 2009).

The first component of a Canadian policy framework for the maritime sector is the normative structure designed to control the operation of ships so as to ensure the safe running of ships and related services. Within this category we can identify safety measures aimed at preventing accidents. Since the mid-1990s this dimension has seen an intensification of the controls and a diversification of risks considered. Most of these changes mirror the transformation internationally through the introduction of new regulations by the International Maritime Organisation. Perhaps a more important change has come about by the International Safety Management Code. This applies to Canadian-flagged as well as to international vessels via port state control. Legislation to prevent intentional harmful acts has come about with the International Ships and Port Facility Security Code. Additional progress has been made with measures to protect the environment. On top of the prevention of accidental pollution, the public framework has turned its attention to operational pollution. In the St. Lawrence Great Lakes this is best exemplified by regulations controlling ballast water discharges to prevent introduction of invasive species, and more recently, rules on ships' air emissions.

The second main dimension of the Canadian policy framework for the maritime sector has been based on creating wealth for the active players of the industry. This time the main influence comes from the United States. The policy framework links the interests of three components of the maritime industry: ship owners, crewmembers and shipyards. To protect these stakeholders from what is judged unfair international competition, the Coastal Trading Act restricts the transportation of goods or passengers between Canadian ports to Canadian-flagged vessels. In turn, conditions to access the national register require that all crewmembers be Canadian citizens and that the vessels be built in Canada. Imports of vessels from other countries is possible, but vessels built in the most efficient shipbuilding countries face an import duty of 25 per cent on the purchase price. This results in a strong segmentation between domestic and international trade. This support of Canadian maritime stakeholders has been strong for decades and remains almost completely intact. Only recently have ships owners publicly voiced their desire to see the 25 per cent import duty be reconsidered, but without questioning the original justification of the whole protectionist framework (for an illustration see Canadian Shipsowners Association 2008).

However more visibility has been given to initiatives to promote and rejuvenate short sea shipping. Echoing arguments developed in Europe and more recently in the United States, campaigns such as the HighwayH2O (HWY H2O 2010) are promoting the sector on the basis that coastal shipping is underused, is not congested, and that it is greener than its truck or rail alternatives. The Government of Quebec has a permanent consultation forum with the industry on this issue. It has developed ongoing programs providing financial support to private firms developing transport services that are not all-road orientated and to shippers electing such services. The other newer development in the public maritime framework is the concern about shortage of skilled crews, resulting in the launch of campaign-type initiatives to address the problem.

The third component of Canadian policy framework for the maritime sector concerns the creation of wealth through the promotion of trade. Here the scope goes largely beyond the maritime sector, as the wealth sought should benefit the whole national economy. Many of the current stakeholders of the maritime sector have been affected by broader policy initiatives such as the opening of the Seaway in 1959, which was built to facilitate access to markets for Canadian industry and to exploit hydro power. More recently, the Canada Marine Act in the late 1990s sought to improve the ability for Canadian ports to support trade, but it was also (or mainly) an operation motivated by the need to reduce the federal deficit. Further current developments concern the renewal of infrastructure through corridor planning.

4. Gateways and Trade Corridors in Canada: The Case of the St. Lawrence Great Lakes System

In order to speak for the private sector, stakeholders of the freight industry active within the St. Lawrence Great Lakes basin formed the Council of the St. Lawrence Great Lakes Trade Corridor. The Council was actually established with the idea of replicating the West Coast approach (see Chapters 10 and 14) and before the actual creation by public authorities of the Ontario-Quebec Continental Gateway, discussed later in this section. The group is representative of the different modal and regional segments, but has deliberately chosen not to include air transportation as its interests were deemed too different. It also clearly focuses on freight mobility and does not consider passenger transportation or individual mobility problems. The secretariat is lodged within the Council for the Economic Development of the St. Lawrence, a maritime oriented advocacy group.

The main activity of the Council of the St. Lawrence Great Lakes Trade Corridor has been to conduct a study to establish the stakeholders' perspective on the corridor's challenges and actions needed to maintain its competitiveness on a 15-20 year horizon, and to identify its priorities in order to prepare advocacies. The study was conducted by the consultants IBI Group. Some 75 industry representatives from about 50 firms in the maritime, rail and road sectors as well as shippers took part.

Released in November 2008, the final report summed up the opinions of the industry participants by identifying eight principal stand alone bottlenecks. Five of these treated the single issue of ships' access and water depths (Groupe IBI 2008):

- Limited water depth in access channels (Quebec City can offer a 15m depth but a section just upriver from the city means that this is limited by a tidal window; up to Montreal the dredged depth is 11.3m).
- Lack of deep berths (creating a limited port capacity for vessels making full use of available channel depth).
- Limitations of the Seaway system in terms of lock size and channel depth, but also in terms of the length of its operational season (currently around nine months a year).
- Potential reduction of depths in the Montreal-Trois-Rivieres segment as a result of eventual changes in how the International Joint Commission regulates the water flow from the Great Lakes into the St. Lawrence.

Several bottlenecks relating to hinterland links were also selected:

- Congestion within the greater Toronto area threatens all hinterland links that go through the region. Included under this inland access problem is the congested road access to the Port of Montreal.
- The Windsor-Detroit rail link: with the tunnel being operated at full capacity, infrastructure investments will be needed to accommodate growth.

The final bottleneck dealt with a maritime (but not water) issue:

- The limited availability of port land. The lack of space in some of the principal ports of the system limits the possibility of accommodating trade from emerging or growing markets and of developing new installations.

The elements selected by this industry survey and the wording used to express them are very informative regarding the conceptions of the private stakeholders. An academic researcher familiar with the St. Lawrence Great Lakes system will readily observe that there are no fundamentally new elements in this list of issues; the handicaps of restricted water access (only compensated by the all important hinterland access towards the Midwest) have been identified from the early studies looking at containerisation in the region (Charlier 1985; Slack 1989) and restated by researchers (McCalla 1994; O'Keefe 2003; Guy and Alix 2007; Langevin *et al.* 2009). Perhaps only the mention of the issue of port land availability would surprise a familiar observer. The problem has been previously identified, but it seems to have climbed significantly within the priorities.

It is also notable that, although the study was presented as a reflection based not on the traditional modal approach but rather on cargo interests, the resulting vision is heavily maritime orientated. Even if there is an intermodal issue mentioned relating to hinterland access issues, participants did not point out any problems of integration of modal players and operations. Participants did point out how the problems seemed aggravated by uncoordinated governmental actions. But they did not see problems arising from uncoordinated operations of private actors, nor that planning and development within a gateway/corridor perspective could require new business models for the actors active in the transport chain.

The IBI-SODES study (2008) also produced a set of recommendations to overcome the bottlenecks and to ensure the competitiveness of the system is maintained. While it was recognised that there was no simple or single solution for the fundamental problems identified, 15 solutions of different nature and scope were presented (a number greater than the number of bottlenecks):

- First in this list comes a very wide recommendation, that seems from the outside more a statement of objective than a specific solution: 'To make Montreal a hub for international container trades'.
- Seven specific solutions established priorities for infrastructure investments:
  - Build a new rail tunnel under the Detroit River.
  - Upgrade the road/rail crossings with new bridges at 21 locations along the Quebec-Sarnia rail network.
  - Maximise roads links within the greater Toronto area and to access the Port of Montreal.
  - Upgrade dry bulk facilities at the Port of Quebec City.
  - Upgrade liquid bulk facilities at the Port of Quebec City.
  - Build new deep berths at the iron-ore terminal in Sept-Iles.
  - Improve the capacity of niche ports for general cargo handling.
- Two solutions echoes the most cited problem of improving water side access into the system:
  - Make the utilisation of depth available (with more flexible rules allowing wider deeper vessels when the conditions are safe) and conduct dredging.
  - Lengthen the season of the Seaway.
- Five solutions targeted improvements to the administrative environment and range from the very general to specific:
  - Establish general and coherent policies to favour terminal and trade corridor related projects.
  - Establish a public-private initiative to address the overall concerns on the availability of labour and to train and retain qualified workers.
  - Harmonise border crossing procedures at the different levels of government.
  - Minimise user-paid fees paid for maritime services (icebreaking, navigational aids or public ports).
  - Abolish the 25 per cent import duty for ships built outside Canada.

The latter point is a priority that has rallied Canadian ship owners in the last two years. It implies a serious reconsideration of one of the pillars of the long-standing Canadian policy by recommending that shipyards should no longer be included in the cabotage regime that governs the system. It is a very delicate issue, and only recently has it gathered sufficient support to become a consensual position among the shipping community of the St. Lawrence and the Great Lakes. This position could aggravate other issues. For example, if the point is made that the protectionist rationale does not make sense for the requirement to use Canadian-built ships, does it trigger a need to look at the obligation to use Canadian crews or Canadian-flagged vessels? The report simply mentions that the current system has not produced a viable shipbuilding industry in Canada and states, without further demonstration, that it has a serious impact on operation costs of Canadian ships therefore limiting their competitiveness.

Industry participants in the St. Lawrence Great Lakes trade corridors apparently did not modify their historic thinking of what is needed when they approached the situation through the perspective of gateway and trade corridor planning and the specifics of cargo-defined transport chains. Their proposed solutions stick closely to the advocacy agenda of the freight industry of the past years. No structural transformations are proposed, but infrastructure improvements and more efficient administrative dispositions are presented. Solutions are mostly linked to a single transport mode at a time or, when they recognise the interdependency of modes, do not propose ways to develop their integration (with the possible exception of labour-related initiatives). Likewise many of the proposed solutions concern a specific location rather than the whole corridor. There is no question that improvements at a specific link can benefit the whole chain, yet some proposed infrastructures show *a priori* little complementarity. Such is the case of an iron-ore terminal in Sept-Iles, for example. The recommendation about developing general cargo capacity in niche ports even specifies that such projects should be judged based upon the individual activities and needs of the proposing ports, seemingly implying that this should predominate over an evaluation of pertinence at the corridor scale.

The IBI study relays the view of a narrower group of stakeholders than the gateway project defined by federal and provincials authorities over the same geographic space. Figure 12.1 is extracted from the memorandum of understanding between Canadian, Ontario and Quebec governments establishing the creation of the Ontario-Quebec Continental Gateway and Trade Corridor as a formal stand alone initiative. It follows on from the National Policy Framework for Strategic Gateways and Trade Corridors (Transport Canada 2007), which sets out the basis on which the gateway model developed on the West Coast could be used elsewhere. More practically, it is through this framework that significant funds (in the range of $2 billion Canadian) would be made available for projects in the Ontario-Quebec Continental Gateway.

The initial aim of the Continental Gateway project is to conduct a wide encompassing consultation of all stakeholders in order to diagnose the challenges of freight mobility in their own region or sector and establish intervention priorities. This work is currently still in progress, with consultation being led by numerous issue-specific working groups. Its conclusions are not available at the time of writing, so it is not possible to directly compare bottlenecks and solutions with those identified by the IBI process.

The organisational chart of the project (Figure 12.1) defines who is involved and how they should relate or interact with each other within the Continental Gateway structure, effectively representing the relative distribution of power among stakeholders. The first characteristic that stands out from the flow chart – aside from its complexity – is that the top of the pyramid is identified as the ministers themselves. The Strategic Leadership Committee acts as the controller of the initiative thus reaffirming the government-led orientation; the first three co-chairs are the deputy ministers of the each of the three transportation ministries. This structure also includes three strategic advisors drawn from private stakeholders whom are chosen, one each, by the deputy ministers. Currently serving are high executives from the cargo handling industry, trucking and the auto-makers. Below this piloting group is a secretariat that is responsible for the operations. Here, the private sector representatives are absent. The assistant deputy ministers of each transportation ministry are in charge with other members drawn from different public bodies such as economic development, border and customs or foreign affairs agencies of the three governments. This secretariat is responsible for coordinating the issue-

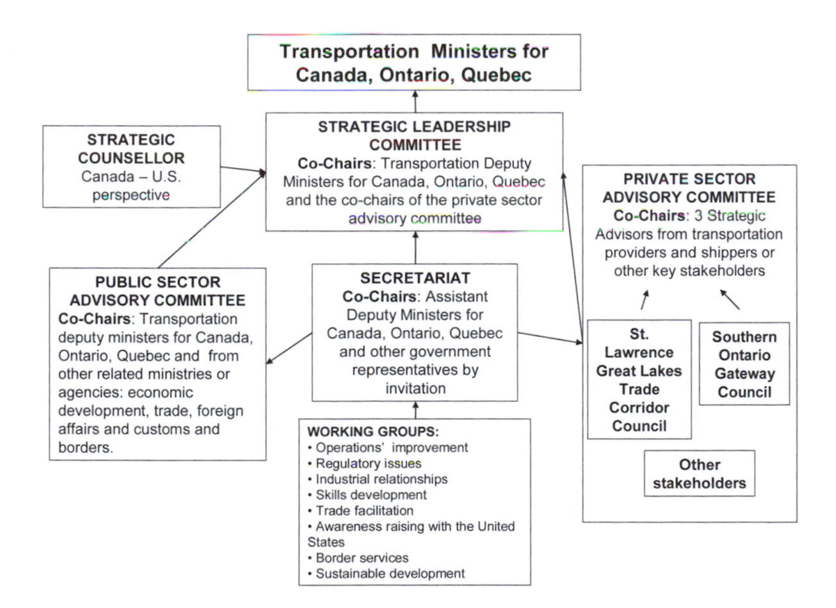

Figure 12.1 Governance Structure of the Ontario-Quebec Continental Gateway and Trade Corridor

Source: Canada-Ontario-Quebec MOU on the development of the Ontario-Quebec Continental Gateway and Trade Corridor and Transport Quebec.

specific working groups in charge of leading the consultation to set the intervention agenda. Private stakeholders are participating directly in those groups.

In parallel with this governance structure, public and private sectors are provided with independent advisory committees. Hierarchically, these advisory committees are above the secretariat relaying their inputs directly to the Strategic Leadership Committee. Visually, the chart suggests that the initiative is conceived to bring together the public and private sectors. Equally noteworthy is the fact that the Council of the St. Lawrence Great Lakes Trade Corridors that contracted the IBI study – itself a wide encompassing consultation and advocacy body with representatives across the freight industry – is just one component of the private sector advisory committee along with, for example, the Southern Ontario Gateway Council.

The material and information used as foundations for this section all come from documents released shortly before the world economy entered into recession that subsequently pulled down the Canadian economy in its wake. As cargo volumes handled at Canadian ports fell in 2009, fears of congestion and insufficient infrastructure have momentarily disappeared. Although threats of congestion had been a major rationale used to justify calls for major investments in the St. Lawrence Greats Lakes basin, it is remarkable to observe the extent to which the maritime industry of this region has stuck to its boom time evaluations and maintained at the top of its priority list the importance of bringing the Ontario-Quebec Continental Gateway and Trade Corridor to a level comparable with its West Coast counterpart. An illustration of this is found in the 2009 Quebec Maritime Day. Held annually, this event is an official occasion for the maritime industry to meet with the elected officials of all parties in Quebec to present their issues and claims to the parliamentarians. A single voice message was sent this year promoting the corridor initiative in accordance to the early priorities, despite the economic perturbations.

5. Challenges of Multi-Goal Policies

Even a partial overview as presented in the previous section illustrates the main difficulties associated with a far reaching initiative that calls for a new approach to transportation planning. We regroup these challenges into three main categories:

- *Policy making with multiple actors (including multiple decision-makers)* – More levels to coordinate enlarges the direct involvement of stakeholders and makes for a heavy structure. The Ontario-Quebec Continental Gateway certainly presents a complex governance structure. A major concern in designing this governance structure appears to be the sharing and circulation of information among the three governments involved and through the different agencies of these governments. This is probably wise and shows an appreciation of this challenging situation from the authorities. Paradoxically, it is likely to make the governing structure heavier. Public servants we interviewed commented that in long careers dealing with complex issues cluttered by provincial-federal duplications typical of Canadian policy, the Quebec-Ontario Continental Gateway has the heaviest consultation and management structure they have ever seen. Inevitably, this leads to slow progress. Fears have been expressed that the first important deadlines fixed for working groups to establish a first hierarchy of investment needs for the summer of 2009 would not be meet. Such a delay may not be very significant or unusual given the scope of the project, but it could eventually open a door for direct political allocation of the significant sums linked to the gateway initiative through the Building Canada Fund. This is a real risk as the economic crisis persists. If significant productivity gains can be obtained by integrating the inputs of all levels of public authorities and private stakeholders of all sectors not only at the diagnostic stage but at the stage of designing responses, this very integration requires a complex governance. This challenging management situation may easily lead to administrative inefficiencies. In response to this risk, the dominant stakeholder may be tempted to impose stricter control of information or impose unilateral decision making. As the strategic orientations that should guide the future action plan are expected to be released in the first half of 2010, one could ask if the economic situation adds pressure to complete the consultation process rapidly, despite its complexity, in order to be able to make investments. Private and provincial stakeholders interviewed observed that despite the fact that they were being consulted, it was clear to them that the Ontario-Quebec Continental Gateway was a federal government led affair with decisions on its orientations clearly taken in Ottawa.

- *Policy making for multiple sectors* – Integrating different modal, cargo-based and regional segments of the transport industry raises governance difficulties because they involve more than one jurisdiction. There are also additional difficulties. In order to rise above listing everyone's specific problems, a common understanding of the issues must be built. To add value to the individual consultation of all stakeholders, an integrated approach must be able to build from everyone's specific problems and develop solutions that would not be reachable from individual standpoints. In other words, a common comprehension of the problem must be developed. Such an understanding would enable each stakeholder to relate to its own difficulties, but also make sense of others' positions and capacities. This may sound like a tautological definition of what an integrated policy approach should be, but in terms of managing the process, achieving a fine and truly common understanding is not a simple matter. The identification of the bottlenecks and proposed solutions in the study conducted for the Council of the St. Lawrence Great Lakes Trade Corridor suggests clearly that this stage has not been reached at the moment. Stakeholders have simply used the opportunity to put forward their specific sector-based or regional issues, issues they have been putting forward before, and are maintaining through the economic crisis.

Closely linked to this issue of common vision are the side effects of competition for public funds. Stakeholders who take part in an initiative such as the Ontario-Quebec Continental Gateway are making an honest contribution towards improving the whole system. But they are also working for their own interests, as it is their responsibility to do. From its very beginnings the Ontario-Quebec Continental Gateway has been presented as a means to allocate in the best possible way the two billion dollars of the Build Canada Fund. It seems clear that many of the stakeholders have thought out their interventions within the initiative and seen that funding would be well spent on their favoured projects. One middle-management of an important port of the system stated bluntly while presenting costly terminal upgrading projects: 'Right now we have the gateway initiative, you know it's a big infrastructure fund, so we're detailing our plans to be ready to present them – just like all the other ports are doing'. While this is not an abnormal or undesirable reaction, it can be asked if the gateway initiative does not harm the integration of the different regional and modal links by spurring competition for public financing. Interestingly, this appears to be even more the case now because of the difficult economic situation. It is interesting to note that there is now a common confusion between the Build Canada Fund that finances gateways projects and the Infrastructure Stimulus Fund which is designed as a supplementary support to spur economic activity with ready to go projects.

- *From adding goals and expectations to arbitraging policy objectives* This challenge echoes the previous two in underlining that to generate added value, an integrated policy initiative must produce more than a list of all the points to consider. Such uncoordinated enumerations can be achieved more efficiently with successive individual consultations. For the result to be superior to the sum of all its components, an integrated initiative must identify priorities that reflect more than stakeholders' relative power and their capacity to push their interests up the priority list. This calls for commonly shared principles allowing a rational evaluation of the respective importance of the multiple objectives and the value of potential measures to achieve these goals. It is the very nature of complex policy issues to give rise to many, and perhaps conflicting, interests. Yet in order to produce measures capable of achieving an acceptable level of effectiveness, there is a need for binding principles on how the objectives interact. This is the foundation of an arbitration of objectives necessary to maintain minimal cohesion among the stakeholders to secure their support. Indications presented here suggest clearly that the various consultations and initiatives around the St. Lawrence Great Lakes system have yet to reach this stage. A comprehensive inventory of stakeholders, issues and solutions is well underway, but it has not produced new means to lead a reorganisation of all these elements towards a fundamentally more integrated system.

6. Conclusion

Wide encompassing policy initiatives such as the current gateway and trade corridor projects within the St. Lawrence Great Lakes system are complex and challenging to manage. To be legitimate, it has to produce more than the simple addition of different stakeholders perspectives. Currently the initiatives have been beneficial to: (i) restate the crucial role of intermodality for the competitiveness of the corridor and the necessity to invest in its infrastructures; and (ii) provide a wide consultation with industry representatives that has given them the occasion to voice the problems and needs at the level of their own links in the chain. Initially begun during a period of strong economic growth, industry appears to have maintained its strong support for the approach during the present economic crisis. Yet the Canadian gateway and corridor approach has not up until now identified means to collaborate and coordinate that would take the integration of the system to new levels. For example, there have been no considerations of the enduring difficult relationships between rail and port sectors; there are no frameworks for a legitimate public intervention in the competition between modes of transport or principles to alter it; and there has been only a limited administrative treatment of ways to promote short sea shipping. Without progress on these considerations, it will be difficult to foresee that the gateway and trade corridor policy will fundamentally add value to the public framework for the maritime sector.

Although there are a multitude of detailed factors that can be built into public management and policy making involving multifaceted issues, this analysis points out that what has taken place so far has little more than symbolic importance. We argue that a set of common views of what constitutes the main dimensions of an issue and how they are interrelated is important not symbolically but in practical terms. They are essential to the legitimacy of this complex issue. Identifying common views is required to establish a rationale that can arbitrate goals and objectives that can be accepted as sound by a majority. For these reasons, stakeholders' values and the conceptualisations of how they are best supported should be central points in coordinating efforts. Our arguments enjoy support from the field of strategic management, where the importance of an organisation's mission and vision is widely recognised. A number of studies have shown the central influence of individual or family leadership in the historical development of liner shipping (Slack and Frémont 2009; Frémont 2007a; Alix *et al*. 1999). Important contributions in political science suggest that representations are at the heart of change in policy as they embody an arbitration between objectives that policies which aim to induce change must tackle (Muller 2005, 2009). Freight transportation research has made very limited exploration of such conceptions of policy. We suggest this would be useful given the global context where intermodal, as well as interregional and inter-jurisdictional, perspectives are increasingly brought into transportation policy and planning. Developing an understanding of the representations underlying the power plays in transportation policy becomes even more pertinent in difficult economic times, as competition for public support is heightened and governments face increased pressure to act promptly to stimulate the economy.

Acknowledgements

The analysis presented in this chapter is derived from research conducted with the financial participation of Transport Quebec and the St. Lawrence Economic Development Council. The authors are thankful for their support, however the views expressed here do not necessarily represent the positions of these organisations.

Chapter 13
Perspectives on Integrated Container Transport: The Canadian Example

Robert J. McCalla

1. Introduction

The term "integration" has many applications. It applies to mathematics and calculus; it has a political connotation as in bringing about cooperation among sovereign states; it applies to the American civil rights movement. It also applies to transportation. It is this latter application that is of interest here especially as it applies to container transport and the various players involved in gateway and corridor development associated with container freight movement. In order for such development to proceed apace it is essential that integration – cooperation and coordination of component parts into a unified system – be operational. As a way to introduce integration in this context I take my lead from two academic articles: Hull (2005) and Hesse and Rodrigue (2004). The former discusses integration related to transportation planning; the second places integration of transportation and logistics into the spectrum of supply chain management. In both cases integration implies coordination and cooperation in which various users – stakeholders – work together for the benefit of the whole, however defined.

According to Hull (2005: 318), integration in a transportation planning context can be discussed in a number of ways. For her, there are two dimensional axes to be considered: horizontal *vs.* vertical. The former involves sectoral integration in which a (local) level of government and various transport service providers come together to provide coordinated public policy and its delivery. The vertical dimension involves inter-government policy formation and implementation. Not only are there different axial dimensions to integration, there are different relational components of transportation to be integrated: transport measures and land use planning policies; transport measures and policies for the environment; and transport measures and policies for education, health and wealth creation (Hull 2005: 320). It is little wonder then that, when public hearings are held reviewing transportation effectiveness and policy related thereto, many different perspectives get aired. There is not just one way to view integration in transportation. This point is fortified by Potter and Skinner's (2000) four nested levels of transport integration moving from the lowest level of functional or modal integration through transport and land use planning integration to social integration to the highest and holistic level of environmental, economic and transport policy integration. As a consequence, different people will interpret the meaning of transport integration. "Much is dependent on what is integrated with what" (Potter and Skinner 2000: 286).

Whereas Hull's perspective comes from integrated transport planning, Potter and Skinner's 2nd level, Hesse and Rodrigue see an integrated transport system more in an operational sense, the 1st level of Potter and Skinner. They see transport as part of physical distribution which, with the other two components of logistics – materials management and strategic planning – makes up supply chain management. Much coordination among the components is required for the smooth functioning of the supply chain. In this light, a transport mode and its operation has impact beyond itself. Thus, when people talk about transportation and its effectiveness they often go beyond discussing the mode; they consider the mode in the broader context of an integrated transport system and the issues at play to enhance the effectiveness of the supply chains in which the transport mode participates.

Both viewpoints of the interpretation of integrated transport expressed by Hull and Hesse and Rodrigue can be seen in the recent hearings (April 2006–June 2008) on container transportation in Canada undertaken by the Canadian government's Senate Committee on Transportation and Communication. The evidence heard at the hearings and subsequent report (Senate of Canada 2008)[1] enable us to see the different ways different players express integration in transportation. The differences mirror the complex relationships transportation – in this case freight transport related to containers – has not only with its component parts but also with other connected sectors, including economic, social and environmental. The issues expressed vary by who is making the statement, where it is being made and in what context integration is viewed.

At the time of the hearing's commencement (June 2006), world containerisation was experiencing unprecedented growth. As de Monie *et al.* show in Chapter 2, annual double digit growth rates in the world container throughputs were the norm, averaging 11 per cent per annum between 2000 and 2006. Canadian container port throughputs followed this trend, although they were not quite as high with an annual average growth of closer to 7 per cent (derived from information contained in *Containerisation International Yearbook*, various years). By the time the Committee had finished its deliberations and the report was released, the world and Canada was on the brink of world recession which would see negative economic growth and corresponding decline in world and Canadian container handling. These declines, though, and the subsequent impact on freight rates, shipping routes, port activity and inland transportation as discussed in Chapter 3 by Notteboom *et al.* do not negate the conclusions drawn by the report. In fact, they make the work of the Committee more meaningful. Integration of all stakeholders in the supply chain is more of a requirement in times of trouble – some would even say crisis – than in times of prosperity and increase.

1 Subsequently referred to in the chapter as Senate Report 2008.

The remainder of the chapter outlines and interprets the Senate's hearings and report with a concluding statement on what can be learned from such an analysis *vis à vis* both Hull's and Hesse and Rodrigue's perspectives on transportation and integration. The chapter ends with a comment on the efficacy of the report's main recommendation.

2. About the Senate Review and Findings

On 11 May, 2006 the Canadian Senate passed a motion:

> that the Standing Senate Committee on Transport and Communications be authorized to examine and report on current and potential future containerized freight traffic handled at, and major inbound and outbound markets served by, Canada's Pacific Gateway container ports, East coast container ports, and Central container ports, and current and appropriate future policies relating thereto.[2]

Thus began the most comprehensive Canadian government review of containerized freight transportation in history.

The hearings and the subsequent report are an example of a national, "top down" approach to understanding a perceived problem and making policy recommendations at a high government level. The hearings were not driven by grass roots advocates arguing that national policies were needed. That being said, the entire undertaking was concerned to hear what stakeholders in the business of container transportation thought about the operations of the industry and listen to ideas of how a national government could support the industry through policy development at whatever level of transport integration was appropriate.

Evidence meetings began on 13 June 2006 and ended on 2 April 2008. In total the Committee heard evidence from 68 presenters at 26 sessions in four different Canadian centres – Vancouver, Ottawa, Montreal and Halifax. The final report entitled *Time for a New National Vision: Opportunities and Constraints for Canada in the Global Movement of Goods* was released on 10 June 2008 (Senate Report 2008).

Of interest to the present chapter, and keeping to the theme of this book, is how the term "integration" was used in association with container transportation. The hearings and the final report did not have as a stated mandate to report on the state of integration within the container transport system. However, by the nature of what was discussed the term "integration" used in various parts of speech – noun, adjective, verb – appeared often in transcripts, and the concept was a common thread throughout the hearings. Of interest here is how the term was presented and what was said about it. Did all presenters have a common view of integrated

---

2 Available at: http://www.parl.gc.ca/39/1/parlbus/commbus/senate/Com-e/tran-e/02or-e.htm?Language=E&Parl=39&Ses=1&comm_id=19 [accessed: 11 February 2010].

Table 13.1 Presenters and Their Affiliation

Government (No. = 9)
Federal: Canada Border Services Agency; Finance Canada; Transport Canada *Provincial*: Nova Scotia Department of Transportation and Public Works *Municipal*: Delta, BC; Moose Jaw, SK; Prince George, BC; Prince Rupert, BC; Saskatoon, SK
Quasi or Non Government Organizations (No. = 15)
Arctic Gateway Council; Asia Pacific Foundation of Canada; Atlantic Canada Opportunities Agency; Atlantic Institute of Market Studies; Atlantic Provinces Economic Council; Greater Halifax Partnership; Greater Vancouver Gateway Council; Halifax Gateway Council; Nova Scotia College of Art and Design; Nunavut Tunngavik Inc.; Prairie to Ports Gateway and Inland Port; Regina Regional Economic Development Authority; Saskatchewan AgriVision Corp.; Southern Ontario Gateway Council; Western Transportation Advisory Council
Transport Industry Providers or Associations (No. = 31)
Ports and Terminals (15): DP World Vancouver; Fraser River Port Authority; Fraser Surrey Docks; Halifax Port Authority; Laurentian Energy; Maher Terminals; Melford International Terminals; Montreal Gateway Terminals Partnership; Montreal Port Authority; Prince Rupert Grain Ltd.; Prince Rupert Port Authority; Ridley Terminals; Termont Terminals; TSI Terminal Systems; Vancouver Port Authority *Trucking (7)*: Association du Camionnage du Québec; BC Trucking Association; Canadian Trucking Alliance; Consolidated Fastfreight; Robert Transport Inc.; Vancouver Truckers Association; West Coast Container Freight Handlers Assoc. *Rail (3)*: CN; CP; Railway Association of Canada *Marine Shipping (3)*: Quebec Shortsea Shipping Roundtable; Shipping Federation of Canada; St. Lawrence Seaway Mgmt. Corp. *Unions (2)*: ILWA (Vancouver); ILWA (Halifax) *Transit (1)*: Translink (Vancouver)
Shippers or Shipper Associations (No. = 5)
Canadian Manufacturers and Exporters Association; Canadian Tire Corporation; Loblaws; Palliser Furniture Ltd.; Pulse Canada
Individuals or Consultants (No. = 8)
Mary Brooks, Dalhousie University; Douglas Campbell, Campbell Agri Business Strategies; Teodore Crainic, UQAM; David Colledge, Colledge Transportation Consulting; John Kosier; Michael Ircha, University of New Brunswick; Barry Prentice, University of Manitoba; John Vickerman, TranSystems

Source: Senate of Canada 2008.

container transportation in terms of whom or what was involved and the benefits of such a policy and practice? If not, what were the different perspectives? Did different perspectives have a bias in terms of the different supply chain actors, whether they came from the private sector or public sector, or in what area of the country they lived and their activities were based? These are the type of questions addressed here. But before proceeding to an analysis of the presentations and their representations on integration it is necessary to give some details on who made presentations to the Committee. Table 13.1 gives such a summary.

As is to be expected, the largest group of presenters (31) belonged to transportation industry providers or associations. Representations were made from the three modes – rail, trucking, and marine transport – but the greatest representation was from ports and port terminal operators. In total, five ports presented their viewpoint to the Committee: Vancouver, Fraser Port, Prince Rupert, Montreal, and Halifax. These ports together account for almost all the international containers handled at Canadian ports. Terminal operators in these ports plus operators from Canso Superport (Melford Terminals) and Sydney, Nova Scotia (Laurentian Energy) also appeared. The latter two operators are hopeful to build new facilities in their respective ports on the Atlantic coast. The federal government was represented mainly through Transport Canada but Finance Canada and the Canada Border Services Agency also appeared. Only one provincial government, Nova Scotia, made a representation. Five municipal governments appeared. Many quasi- and non-governmental organizations appeared. These organizations are mainly involved in analyzing and promoting regional economic development in the areas in which they are located. Only three private sector shippers gave evidence to the Committee – Canadian Tire, Loblaws and Palliser Furniture – but the Canadian Manufacturers and Exporters Association and Pulse Canada also made presentations. Eight individuals appeared, three of whom were academics. Surprisingly, the other components of the logistics industry – freight forwarders, customs brokers, warehouse operators, distribution centres – were not well represented. Only Consolidated Fastfreight would qualify under this category. Also surprisingly, there was no representation from a container shipping line to the Committee. In sum, the ports, inland transportation companies, economic development agencies, the federal government and individuals with professional interests in Canadian container transportation all made presentations to the Committee.

The geographical representation spanned the country including the Arctic, but for the most part the majority of the representation came from western and eastern coastal areas. It was surprising to see such little government and non-government representation from Ontario; only the Southern Ontario Gateway Council made a presentation. This is surprising because approximately 40 per cent of the Canadian population and domestic product is produced in Ontario, and as such forms the destination and is the origin for almost half of the container traffic in Canada.

2.1 What was Said

The focus of this chapter is integration as applied to container transportation in Canada and how this was expressed to the Committee. Based on thematic coding on the word "integrate" or "integration" applied to the verbatim text of hearing transcripts and the final report, statements were isolated that form the basis of the analysis presented here. Although the presenters were all directing their remarks to container transportation, there were various themes expressed as to what integration actually meant; that is, what aspect(s) of container transportation was (were) involved to create integration or what was needed to work with to create an integrated container transportation system. Some saw integration as a policy issue; some saw it in an operational light; some saw it as a local issue; some saw it nationally and internationally. In the end there is no one way that all presenters expressed themselves about integration of container transportation in Canada.

Five themes of integration in container transportation could be identified:

1. Related to transportation policy
2. Related to supply/logistics' chains
3. Related to intermodality and transport systems
4. Related to social and/or environmental concerns
5. Related to economic development.

Each of these themes is discussed below with emphasis on the first three.

2.1.1 Transport policy Integrated container transportation as a policy issue was expressed by the Minister of Transportation in the statement: "The Government of Canada … has been working in partnership with key public and private sector partners to address the need for more *integrated approaches* to transportation policy, planning and investment" (Lawrence Cannon, Minister of Transport, Infrastructure and Communities, 4 October 2006).[3] In this context, integration is an approach to better decision making. Taking an integrated approach to transport policy development is not new. The European Commission's 2001 white paper on transport policy stressed the need to have a common transport policy not only for the betterment of the economy but also the environment and society at large (European Commission 2001). Furthermore the EU's 2007 communiqué on the European freight transport agenda states that "To improve the efficiency and sustainability of freight transport, the authorities have to create the appropriate framework conditions (i.e. policy) and support the trend towards commodality and sustainability" (European Commission 2007). At another geographical scale, recently the State of Victoria, Australia, enunciated that it would be bringing forward legislation that "will provide a clear policy framework for transport decision-making to ensure that the actions and activities of transport bodies are complementary and work towards the delivery of a common vision." (State of Victoria 2009).

Minister Cannon was not alone in his reference of integration as a policy issue, but more specifically as a policy issue related to trade. "This issue of national transportation policy is not one of how we regulate it, but it is how we *integrate* transportation with our trade policy" (Allan Domas, President and CEO, Fraser River Port Authority, 13 March 2007). The concern between transport policy and

3 References to direct testimony heard by the Committee are indicated by person and date. Testimony heard between June 13, 2006 and June 12, 2007 can be found at: http://www.parl.gc.ca/common/Committee_SenProceed.asp?Language=E&parl=39&Ses=1&comm_id=19 [accessed: 11 February 2010]. Testimony heard between 13 November 2007 and June 3, 2008 can be found at: http://www.parl.gc.ca/common/Committee_SenProceed.asp?Language=E&Parl=39&Ses=2&comm_id=19 [accessed: 11 February 2010].

trade is a common one (Lakshmanan *et al.* 2001; UNCTAD 2006; UNCTAD 2007). Both Cannon's and Domas' expressions of integration are "big picture" items that go beyond day to day operations. The Committee picked up the shortcomings of national policy in container transportation in its final report and spoke of the need for better integrated policies related to container transportation and trade. The most pointed statement in this regard was:

> The container system in Canada is made up of a number of components – ships, railroads, trucks, container/marine terminals, information technology and labour. The policy environment in which these various components operate is a patchwork of federal, provincial and municipal jurisdictions. (Senate Report 2008: 8)

The Committee went on to say that it believed "that there is a significant policy role for the federal government to play in further developing this [container transportation] sector. It can provide the leadership role necessary to coordinate efforts among all the players in the container transportation industry" (Senate Report 2008: 9). To this end it recommended that an independent National Gateway Council be created to "assist the stakeholders in the container transportation to bring together national and international players in the container transportation system and governments from across the country to enhance communications, bring efficiencies to the system and market Canada's container transportation system to the world" (Senate Report 2008: 13).

2.1.2 Supply/logistics chains Many presenters expressed that container transportation is about supply or logistics' chains and the need for integration or cooperation/coordination within them. Such expressions came from port operators, executives in quasi- or non-government organization and consultants. The most comprehensive expression of what constituted an integrated chain was made by the then-Executive Director of the Port of Montreal speaking about the port:

> We are at the centre of a diversified and *well-integrated* logistical chain: terminal operators, coast guard, dock workers, railway transports, truck drivers, freight forwarders, shipping agents, warehouses, government agencies, longshoremen and checkers, shipping line pilots, shipping agencies [author's emphasis added]. (Dominic Taddeo, 7 February 2007)

These remarks though were specific to the Port of Montreal; they did not reflect a national perspective on such an integrated system. Others were broader in their geographic expression of integrated supply chains or logistics systems: e.g. "gateway and corridor initiatives are largely about expanding our *integration* into global supply chains and logistic systems that more deeply connect North America to Asia, and, in particular, Canada to Asia" (Paul Evans, Co-CEO, Asia Pacific Foundation of Canada, 13 March 2007) and "I have heard calls in the industry for an *integrated* national logistics strategy to make Canada the trade hub of North America" (David

Colledge, Colledge Transportation Consulting, Inc., as an individual, 14 March 2007); and finally "gateways and corridors should be developed as part of a larger piece, a national vision, an *integrated* system with good policies and a means of funding, a grand plan of interdependent supply chains" (Ruth Sol, President, Western Transportation Advisory Council, 14 March 2007).

In response, the Committee acknowledged the changes in global business and its reliance on integrated transportation within global supply chains. It went as far as to say: "We strongly believe that container transportation must be viewed as a system and that each part must function efficiently if the supply chain is to flourish" (Senate Report 2008: 19). But it also recognized that the system in Canada was flawed with port congestion, system unreliability, labour problems, lack of government policy, outstanding aboriginal issues and security concerns. To address these problems the Committee made many recommendations that spoke directly to the mode involved (railroad, truck, short sea) to port operations and to integration among the modes and with United States operations and policies in order to develop an integrated continental supply chain. These various recommendations are shown in Table 13.2.

Table 13.2 Selected Recommendations Regarding Operation and Integration of Supply Chain Actors

Mode	Recommendation re: Operations	Recommendation re: Integration
Railroads		The government examine the reasons for and take appropriate actions to address the lack of consultation between the railroads, shippers and ports regarding service delays.
Trucking	The Federal Government work with the Canadian Council of Motor Transport Administrators (CCMTA) to harmonize trucking regulations across the country.	
Short Sea Shipping		The government support the growth of short sea shipping by negotiating with the US to exempt shortsea container vessels from the [US] harbour maintenance tax; and negotiating multilateral cabotage exemptions for shortsea container shipping operations.
Ports	The government implement the recommendations contained in the 2003 report of the Canada Marine Act review panel which called for the ports to be able to issue tax-free bonds and pay the stipend based on their net, not gross, revenues.	

Source: Senate of Canada 2008.

Cooperation and coordination among supply chain actors to achieve a more integrated transport system is an ongoing focus of many ports, land transport providers and logistics companies. It is the key to better efficiency leading to cost and service advantages. Examples are legion (Brooks *et al.* 2009). As cooperation examples, consider that Rotterdam and Amsterdam now have a merged port data system in order to offer customers a broader range of services for the exchange of data both between them and with the port authorities and Customs. Also, Tacoma and Seattle have been working together on road and rail projects for improved access to port areas. Coordination examples among ports and transportation providers can be found in many ports where they have broadened the scope of their operations to include development of inland terminals, equity investment in intermodal rail and operations of trucking and feeder services (Notteboom 2004a). But for the most part the scale of these examples is local and regional; it is not national, which was the focus of the Senate review.

2.1.3 Intermodality/transport systems/transport networks Just as there were expressions of integration involving supply and logistics chains, so also many spoke of integration related to intermodality involving transport systems. The most succinct way in which integration, intermodality and transportation were linked was expressed as: "transportation is a fluid and extremely integrated system. Intermodality is an *integrated* concept. You cannot talk about intermodality and focus on a single stakeholder, a single mode of transportation" (Teodor Crainic, Université du Québec à Montréal, 15 May 2007). The Port of Montreal also recognized the relationship between integration and intermodality but, again, it confined its perspective to the local area, as seen in: "The Port of Montreal is a formidable gateway for trade, with its *integrated* intermodal system that brings together and develops privileged relationship with all its stakeholders" (Patrice Pelletier, Chief Executive Officer, Montreal Port Authority, 6 February 2008). A narrow geographical perspective on integration and transportation systems was also expressed by the Executive Director of Armateurs du Saint-Laurent and President of the Quebec Shortsea Shipping Roundtable when speaking about a joint federal-provincial assistance program within that province to "*integrate* the various modes within Quebec's transportation system while also taking into account competitiveness, reducing social costs of transport activities and protecting the environment" (Nicole Trépanier, Executive Director, Armateurs du Saint-Laurent, 15 May 2007). This regionalism in integration and transport systems is also expressed by the Executive Director of the Southern Ontario Gateway Council as seen in: "Our mission statement is to achieve excellence in an *integrated* transportation system for the prosperity of Southern Ontario" (John Best, 5 June 2007).

Much of the comments about integration, intermodality and transport systems had a local or regional perspective. This is what could be expected given that the majority of presenters represented or were involved with a local or regional intermodal network. Committee members made comments to that effect and even asked some witnesses their perspectives on a wider geographical focus. For example, Senator Eyton picked up on the local perspective of Kevin Doherty of the Montreal Gateway Terminals Partnership when he said "As a committee out of Ottawa, we are looking at some of the countrywide or national concerns, and one thing we have heard about and thought about a good deal is an integrated, national shipping transportation corridor that really goes from sea to sea. … However, that does not seem to be a factor in your presentation…" (15 May 2007). As well, Senator Zimmer forced a wider geographical perspective on integration and intermodal transport systems in questions asked of witnesses. Two verbatim exchanges are below.

From 8 February 2008 hearings in Halifax:

> *Senator Zimmer*: How important is it that Canada forms part of an integrated North American container transportation system from the perspective of the rest of the world?
>
> *Mr David Chaundy* (Senior Economist, Atlantic Provinces Economic Council): We need to see a stronger engagement by Canada in ensuring that developing security practices and policies in the US do not impede business between the two countries anymore than is necessary. Therefore, there is a role for Canada … The more steps that could be taken to help strengthen that relationship and manage the border and security issues would be an advantage to all North American companies.

From 9 February 2008 hearings in Halifax:

> *Senator Zimmer*: You talked about niches and that we have to all work together. How important would it be for Canada to form part of an integrated North American container transportation system from the perspective of the rest of the world, not just across Canada but also North America?
>
> *Mr Charles Cirtwill* (Acting President, Atlantic Institute for Market Studies): I think that is absolutely central … In terms of the federal government and its responsibility for foreign relations and integrating with the United States in particular, the solutions to many of the US transportation problems lie on this side of the border so I think it is actually critical to have a North American approach to this issue.

The final report of the Committee highlighted the systematic nature of container transportation and the need to have integration of the modes, but it also pointed out that "many of the players seem to work in isolation" and "better planning and coordination throughout the industry" (Senate Report 2008: 9) is required. The isolation can be seen in the local and regional comments expressed by some of the witnesses. The recommendation to create a National Gateway Council (see above) to promote the efficiency of the whole transportation network and market its use to the world speaks to the need to bring the various components of intermodality together.

2.1.4 Social and/or environmental concerns Relatively little was said about integration of container handling and social and/or environmental concerns. Most witnesses discussed operational problems of supply chains and transport providers from a business perspective. Reference has already been made to Trepannier's statement (15 May 2007) that a Quebec integrated transport system would reduce social costs and protect the environment. So also Richard Corfe, CEO of the St. Lawrence Seaway Management Corporation saw that "the social and environmental benefits of shipping along the seaway must be *integrated* into the total cost equation" of Seaway operations (15 May 2007). Finally, David Chaundy, Senior Economist, Atlantic Provinces Economic Council cautioned that:

> There are still policies that do affect the movement of containers that could be looked at … whether we have the right overall strategic framework given issues of increasing concern about the environment and whether we are fully *integrating* environmental concerns into our transportation policy … (8 February 2008).

Although relatively little was said about integration of transportation and environmental impact at the hearings, the final report made four specific recommendations related to this theme. Three spoke specifically to the trucking industry – to harmonize trucking regulations with the provinces to ensure the adoption of energy efficient technologies, to review capital cost allowance regulations to enable faster write-down of investment in energy efficient trucks, and to increase funding for more grade separations between trucks and rail to minimize congestion and reduce idling time. Of these only the third related to integration of transportation; the other two were mode specific. The fourth recommendation was a general one having to do with speeding up the environmental review process for new transportation infrastructure.

As for integration of transportation and social concerns, the final report made no recommendations, and in fact, only once in the final report used the word "social", when speaking of the need to heighten the importance of transportation in the Canadian economy "because a robust transportation economy would help finance other national priorities such as healthcare, education and *social* programs" (Senate Report 2008: 72).

2.1.5 Economic development Although there is a tacit relationship between transportation and the health and prosperity of an economy, relatively little was said about such a relationship by witnesses to the hearings. This is even more surprising given that at least four presenters were from organizations – Atlantic Provinces Economic Council, Nunavut Tunngavik Inc., Prince Rupert and Port

Edward Economic Development Corporation, and the Regina Regional Economic Development Authority – whose mandate, in whole or part, is the economic development of the areas which they represent. Also, the private sector witnesses – Canadian Tire, Loblaws, Palliser, Canadian Manufacturers and Exporters Association and Pulse Canada – represent businesses whose responsibility is to make a profit and are heavily dependent on transportation to do so. The only direct statement linking an integrated transportation system with economic development was made by the Executive Director of the Southern Ontario Gateway Council (quoted above). Comment has already been made on the narrow geographical focus of the statement.

On the other hand, the final report of the Committee certainly acknowledged the close parallel between transportation and economy. Statements such as "The federal and provincial governments, regional interest groups, supply chain players and other stakeholders recognize that substantial economic benefits can be derived from an efficient container transportation system" (Senate Report 2008: 8) and "If Canada can maintain or increase its share of North American container trade, there is a huge payoff in terms of economic growth, jobs and improved standard of living for all Canadians" (Senate Report 2008: 17) are indicative of the relationship. That being said, no specific recommendation was made about improving the relationship, although it could be argued that the entire report had as its overriding mandate (unsaid) of improving the policy and operations of an integrated container transportation system for the benefit of the Canadian economy.

3. Discussion

Many viewpoints on integration and transportation were expressed to the Committee. They have been broken down into the five themes discussed above, but overwhelmingly the views expressed by the witnesses related to transport policy, supply/logistics' chains and transport systems, and so also did the final report.

The transport policy viewpoints relate well to Hull's vertical dimension of transport integration, that is, to inter-government policy creation involving different tiers of government. These tiers may be, in the Canadian context, municipal and/or provincial, but also they involve international, especially United States, policies. A major recommendation of the Report, to create a National Gateway Council, spoke to the necessity of bringing together not only transport providers but also "governments from across the country" to create efficiencies in container handling.

The supply/logistics chain views all relate to Hesse and Rodrigue's theme that an integrated transport system is part of supply chain management, but the geographical scale at which such a system operates was not consistent among the presenters. Some saw it parochially; others were broader in the viewpoint. The parochial view fits into Hull's horizontal dimension of transport integration involving local government and local service providers, but from a national government perspective as represented in the Senate hearings such a view is not useful. National governments can only provide policy which applies to the large geographical scale. They will appreciate that issues exist at local levels, but their mitigation must be handled locally. This is why a National Gateway Council as proposed by the Committee must involve representations from governments below the national level; thus, Hull's vertical dimension comes to the fore.

The views expressed by witnesses on intermodal/transport systems/transport networks relate to the physical distribution function of Hesse and Rodrigue's theme of logistics and integration. Again, many of the perspectives expressed were local and the Committee attempted to get witnesses to change from the horizontal to the vertical perspective of operations. It is not surprising that presenters had a local perspective on the operation of the physical distribution. Given that almost half of the presentations came from ports and terminal operators who are caught up in their local concerns, and know very well their local operations, it is difficult for them to take a wider view of issues, especially a view that incorporates vertical thinking of governments and transport providers coming together in an integrated fashion.

As for Hull's relational components of transportation in the areas of transport planning, impact on environment and social responsibilities, relatively little was said by the presenters. This is understandable given that the thrust of the hearings was to make policy recommendations regarding the movement of containers through Canadian ports for the benefit of the Canadian economy. Planning and environmental and social responsibility took a back seat to reporting on difficulties of operation and where improvements could be made to make the system more efficient. The witnesses mainly brought a business operations perspective to their presentations. The report itself, though, did represent the need for integrated transportation in planning, and its importance in environmental and social impact, much more than the presenters. Again, this is understandable given that the Committee had a national perspective to represent – one that went beyond the business perspective of container handling. But still the emphasis in the report is not in relational components of transportation beyond the actual modal and logistics operations. Rather, it focused on Potter and Skinner's lowest level of transport integration, i.e. on functional and modal integration at the operational level.

4. Conclusion

Integration as a concept has many connotations. Even when confined to transportation, integration is interpreted in different ways. These differences are certainly seen in the testimony of witnesses to the Senate review of container handling at Canadian ports. This chapter has summarized those viewpoints under five themes, and also relative to two academic papers which are useful to give context to the discussion. From an integration perspective, no one theme dominated the presentations.

The report of the Committee represented well the various viewpoints expressed. The Committee certainly grasped that the integration of the various facets of container transportation in Canada was essential for smoother operation of the system as seen in: "We believe that communication and coordination among all players in the system are extremely important to achieving a seamless transportation network. Without this, stakeholders are not participating to the fullest in an *integrated system* but rather, they are operating in their own particular silos" (Senate Report 2008: 75). The lack of coordination and operations in isolation was seen as failings of the present system. To overcome such deficiencies, the "overarching recommendation" (Senate Report 2008: 80) of the final report is the creation of a National Gateway Council to bring stakeholders – governments and logistics providers – together to enhance communications in order to bring about efficiencies and market Canada's container transportation system to the world. By so doing the perspectives of both Hull and Hesse and Rodrigue would be represented. Not only would Hull's horizontal and vertical dimensions of integrated transportation planning be addressed, but they would be focused on Hesse and Rodrigue's integrated transportation within the context of supply chain management. Although recommending a National Gateway Council would address, conceptually, the perspectives of Hull and Hesse and Rodrigue, its creation would bring operational challenges. Having interested parties coming together at a national level to air grievances and suggest remedies is only effective if what is agreed to is accepted and implemented at a local level. In the Canadian context where responsibility for transportation is held at all three levels of government – federal, provincial and municipal – it is difficult to get cooperation and coordination, that is *integration* among the public players. This difficulty is compounded by the private operators who must adhere to the laws and regulations that exist at all three public levels. Such is the challenge of a federated system and a top down approach to systems operations spanning different jurisdictions and interest groups.

Effective operation of national coordinating bodies would seem even more important today as the world struggles to overcome the economic downturn and the future of containerization looks to be very different from the past. It is vital that all governments and logistics providers come together to enhance efficiencies within the transportation system. This is a statement which applies beyond the Canadian context presented here. Needless duplication and uninformed decision making must be overcome, paving the way to a smooth transition when the economy, world trade and container shipping turns from decline to growth.

Chapter 14
Trade Corridors and Gateways: An Evolving National Transportation Plan

Michael C. Ircha

1. Introduction

With its relatively sparse population and immense distances between major urban centers, Canada has always depended on efficient transportation to link its disparate regions together. Indeed the nineteenth century Confederation of Canada was predicated on joining several British colonies with an east-west transcontinental railway. This key transportation system opened the country's interior to global trade, enabling export resources to be sent by rail to distant coastal ports. Developing Canada's east-west transportation corridor continued into the twentieth century with the federal government's creation of the Trans-Canada Airlines (the predecessor of Air Canada), Trans-Canada Highway and Trans-Canada pipeline.

The 1988 US-Canada Free Trade Agreement (FTA) and subsequent 1992 North American Free Trade Agreement (NAFTA), which includes Mexico, affected Canada's traditional east-west trade corridor by redirecting a significant portion of the country's continental trade north-south. For example, by the mid-1990s, Canada's north-south intermodal railway traffic had surpassed east-west domestic movements (McMillan 2006). CN's railway acquisitions in the 1990s reflected this north-south shift by providing rail access through the central US to the Gulf and northern Mexico. CP Rail followed suit with similar northern US railroad acquisitions.

NAFTA's north-south trade orientation led many groups in Canada and the US to focus on the benefits of designated continental trade corridors. This resulted in a plethora of trade corridor promotion organizations throughout the continent. Trade corridors have been defined as "streams of products, services and information moving within and through communities in geographic patterns" (Van Pelt 2003: 6). Another interpretation considers "North American trade corridors as strategies developed by groups of business and municipal (and sometimes state and even federal) government leaders to attract to particular regions some of the increased flow of materials generated by deepening North American economic integration" (Blank 2006: 2).

During the past two decades, governments at all levels paid increasing attention to trade corridors. Their objective was to encourage significant public and private investment in transportation infrastructure and related economic development in designated corridors to facilitate trade and stimulate economic growth. Many proposed north-south trade corridors linked US Interstate Highways with their Canadian counterparts. In 1991, the US envisioned an integrated system of superhighways supported by the Intermodal Surface Transportation Efficiency Act (ISTEA). Politics soon intervened with Congressional members adding their favored routes to designated high priority corridors in subsequent legislation such that these corridors now cross the country in a maze of routings in all directions. As a result, other than some improvements to selected border crossings, little has been achieved. A coherent, rational integrated North American highway system has yet to be achieved (Blank 2006).

Canada's major trade corridor is clearly east-west, supplemented by NAFTA's north-south shift. This traditional east-west corridor serves most of the country's population who live in a linear band within 200 kilometers of the US border. But, despite the geographical advantage of a domestic unidirectional trade corridor, many challenges face Canada's transportation system. These include differing rules, regulations, and taxation regimes among Canada's ten provinces, as well as border crossing issues and similar regulation differences among many US states.

Over the past several decades, efficient and low-cost transportation enabled Canadian trade to become increasingly international by sourcing and manufacturing components and products in lower-cost countries, particularly in Asia. Such economic globalization led trade corridor proponents to consider ports to be key "gateways" connecting their corridors to the global marketplace.

However, as pointed out by Levinson (2008), higher energy prices in the summer of 2008 marked the beginning of a changing global economy. Fuel surcharges drove up transportation costs to the point where global industries considered relocating their sourcing and manufacturing options closer to primary markets. The more recent changing global economic crisis has led to a dramatic decrease in manufacturing and the sale of consumer goods. The rapid contraction of the global economy has resulted in a decline in international trade that has reduced the demand for shipping. This in turn has led to lowered freight rates, shipping overcapacity and diminished port throughputs.

Over the years, the challenges facing Canada's, and indeed, the continental transportation system have led many transportation analysts, politicians, operators and others to call for a national transportation plan or strategy– a plan to rationalize the myriad of conflicting regulations, financial issues and infrastructure deficiencies plaguing Canada's, and the continent's, transportation network. Such a national or continental strategy would depend on collaboration and coordination of many parts of Canada's transportation industry. This chapter explores whether the recent Canadian designation of trade corridors and gateways beginning with the Asia-Pacific Gateway and Corridor Initiative, then the Ontario-Quebec Continental Corridor and Gateway and the Atlantic Gateway may generate a "bottom-up" national strategic transportation plan.

2. Demands for a National Transportation Plan

Over the years, the transportation community has called on the federal government to develop a national transportation plan. The most recent of these demands for a national approach came from Canada's Senate Standing Committee on Transport and Communications and the Canadian Chamber of Commerce. The Senate Committee's report *Time for a New National Vision – Opportunities and Constraints for Canada in the Global Movement of Goods* called for a:

> … National Gateway Council to bring together the players in the container transportation system and governments from across the country to enhance communications, bring efficiencies to the system and market Canada's transportation system to the world (Senate of Canada 2008: 80).

The basis of the Senate Committee's concern for improved collaboration and coordination arose from the many constraints they found in the container transportation sector, such as:

- A need to update port policies including the adoption of key recommendations of the Canada Marine Act Review Panel (several key recommendations were adopted in amendments to the *Canada Marine Act* in the summer of 2008);
- A need for balanced trade in the railroad sector to ensure economic efficiency;
- Current and forecast labor shortages, particularly in the long-haul trucking sector;
- Inadequate use and sharing of electronic data amongst various participants involved in moving containers from origin to destination;
- The long-standing patchwork of differing provincial and federal trucking regulatory regimes;
- Contradictory national and international regulations stifling efforts to develop an efficient short sea shipping regime;
- Over-lengthy environmental assessments and differing federal and provincial approval procedures; and
- A need for priority setting for transportation infrastructure investments.

The Senate Committee's response to these many transportation constraints was to recommend the creation of a National Gateway Council. This Council's goal would be to improve collaboration and coordination, increase infrastructure funding and harmonize regulatory policies.

In a similar vein, the Canadian Chamber of Commerce recently released its transportation review *Moving the Canadian Economy: Four Pillars for a National Transportation Strategy*, which recommended:

> … that the Government of Canada develop and implement a *National Transportation Strategy* for the movement of goods and people. It must ensure that all parts and levels of government share a common vision while working collaboratively toward a common goal … A *National Transportation Strategy* needs to be in place to guide the actions of all government departments, not just the action of one ministry (Canadian Chamber of Commerce 2008: 4).

The Chamber's call for a national transportation strategy echoes the Senate Committee's National Gateway Council. Both groups appear to be seeking a national, centrist, "top-down" approach to address Canada's many transportation challenges.

However, expecting the federal government and its transportation agencies to develop such a centrist solution may well be a thing of the past. With financial cuts and significant staff reductions in recent years, most government departments and agencies are struggling to maintain their limited mandated regulatory responsibilities. As a result, these resource limitations preclude Transport Canada from embarking on ambitious projects including a national transportation strategy.

Beyond these departmental resource constraints, Canada's repatriated *Constitution Act* of 1982 creates constitutional barriers for federal–provincial initiatives. Specific government powers are divided between the federal government and the provinces. Sections 91 (federal) and 92 (provincial) in the *Constitution Act* list the areas in which each jurisdiction can enact laws. From a transportation perspective, the federal government has jurisdiction over inter-provincial and international activities such as air travel, marine shipping and ports, border crossings, class 1 national railways, inter-provincial trade and support for the National Highway system. Provinces have jurisdiction over intra-provincial transportation including highways and short-line railways. To add to this jurisdictional mix, the provinces' municipalities and their local transportation interests along with those of private operators and users all have an interest in a national transportation strategy or plan. This broad mix of federal, provincial, municipal and private transportation interests creates difficulties in devising a coherent and acceptable national transportation vision.

Since Canada's Confederation in 1867, provincial responses to transportation infrastructure, regulations, and taxation issues have varied across the country. These variations led to a myriad of differing standards, rules, regulations and fiscal regimes. This happened so much so that the Canadian Chamber of Commerce argued that "our transportation regulatory environment, consisting of inefficient tax and operating requirements and split responsibilities between levels of government, requires modernization" (Canadian Chamber of Commerce 2008: 1).

Canadian provinces have different truck weights and dimensions, varied taxes on cross-province trucking and rail movements, and subsidy levels (for example, assessments for property taxes). On a North American scale, the US has a similar myriad of state transportation regulations and fiscal regimes. To further complicate things, Canada's federal government handles its transportation systems differently than the US. For example, Canada Port Authorities are required to contribute an annual stipend to the federal government based on their gross revenues along with paying property taxes or grants in lieu of taxes to their host municipalities. In the US, port authorities either pay little tax or have taxing authority to generate tax revenues from their host municipalities. US ports have the ability to raise capital by issuing tax exempt revenue bonds while Canadian ports are expected to borrow capital funds from private financial institutions, and until recently, were prohibited from accessing any federal funding (Ircha 2001). Recent amendments to the *Canada Marine Act* now allow Canada Port Authorities to seek federal funds for specific capital needs related to infrastructure, security and environmental sustainability. Despite this amendment, the financial flexibility of Canadian ports is still a long way from the broad-based fund raising abilities of their competing US counterparts.

A major problem stemming from the *Constitution Act's* divided jurisdiction is the lack of a cohesive approach to multi-modal transport. Provincial governments, with their key transportation responsibility being roads and highways, tend to pay short shrift to the needs of other transport modes (as these are considered to be federal concerns) (McMillan 2006).

However, over time these challenges are being tackled. The annual Council of Ministers and Deputy Ministers Responsible for Transportation and Highway Safety has considered multi-jurisdictional issues aimed at achieving a rational, national approach. For example, the Council encouraged the collaboration and coordination of the many players and jurisdictions in the national system in their policy paper "A Partnering for the Future: A Transportation Vision for Canada" (Council of Ministers 2002). The Council also has a long-standing Task Force on Vehicle Weights and Dimensions Policy. A recent output of this Task Force is "Harmonization of Transportation Policies and Regulations: Context, Progress and Initiatives in the Motor Carrier Sector" (Council of Ministers 2008). A similar, but more senior, Council of First Ministers (the Prime Minister and provincial premiers) meets annually to discuss and deal with issues of national significance. The new global economic context led Canada's First Ministers to take steps to remove persistent domestic trade barriers to the inter-provincial movement of goods and people (labor and professional mobility).

Despite these valiant efforts at collaboration and coordination, many constitutional differences remain. These differences, coupled with significant resource constraints in Transport Canada and other government departments, prevent the creation of a comprehensive, centrist, top-down national transportation plan. In today's global economic context with numerous stimulus programs, such a national plan is essential to guide strategic transportation infrastructure investments.

The gateway and trade corridor concept focusing on regional economic development reflects a linear version of earlier growth pole theories posited by

Perroux (1950) and others (Buttler 1975; Lo and Salih 1978). Many economists applied growth poles or industrial clusters in a regional geographic sense (Friedmann 1966; Parr 1999). In France, growth poles were used to define key ports coupled with linear industrial corridors linking them to major urban centers such as the Le Havre-Rouen-Paris axis (Tuppen 1983). Growth poles purportedly supported strategic public and private investments in targeted areas to achieve optimal results. Designating linear trade corridors and key port gateways in Canada serves a similar purpose by focusing public and private transportation investment within strategic geographic regions in support of international trade.

3. An Alternative "Bottom-Up" National Transportation Plan

Just as a centrist top-down strategy for developing a national transportation plan seems impossible, an alternative, more "bottom-up" approach appears to be evolving. In recent years, the federal government's transportation focus shifted to encouraging designated trade corridors and strategic gateways. This more regionally and locally based strategy may offer a key to a collaborative, coordinated and comprehensive national transportation plan.

At an international level, the concept of north-south trade corridors flourished with earlier "Cascadia" initiatives on the West Coast and the more recent "Atlantica" discussions on the East Coast. These initiatives focused on cross-border regional economic development. The Atlantica initiative proposed a trade corridor linking Atlantic Canada with several New England States and northern New York (Gudeon 2005). Cross border links between Canada and the US are strong in many regions. For example, there are numerous socio-cultural linkages between the Maritime Provinces and New England states. In the past, during periods of economic depression, Maritimers tended to travel south seeking employment. Such north-south social, cultural and economic integration reinforces cross-border trade opportunities (PRI 2008).

In Western Canada, a coalition of private and public sector transportation interests collaborated over a decade ago to establish the Vancouver Gateway Council to rationalize transportation infrastructure services and investments to meet growing Asian trade demands. This coalition evolved into Canada's first successful trade corridor and gateway when the federal government launched the Asia-Pacific Gateway and Corridor Initiative (APGCI) in October 2006. A key factor in the federal government's support for the APGCI "was the extent to which a stakeholder-driven consensus had taken shape over a number of years" (Transport Canada 2007: 12). APGCI focuses on the British Columbia and Prairie hinterlands of ports in Vancouver and Prince Rupert and has received over $860 million in federal funding supplementing provincial, municipal and private support to develop and enhance essential transportation infrastructure. In total, the public sector is:

> … investing about $5 billion over five years and attracting private investment of about the same amount for an ambitious series of infrastructure projects focused on ports, rail and roads intended to create a seamless, multi-modal network that can move people and goods across the Pacific and throughout North America more efficiently (Evans 2008: 94).

APGCI's aim is to reduce congestion and ease the flow of goods in and out of the major Canadian ports of Vancouver and Prince Rupert. As part of a long-term rationalization strategy, on January 1, 2008, the federal government amalgamated three Vancouver area port authorities (Vancouver, Fraser River and North Fraser River) into one unit, now known as Port Metro Vancouver (Houston 2008). Amalgamation was aimed at making effective and efficient use of Vancouver's port resources on land and water.

Following APGCI's success in attracting substantial transportation infrastructure funding, other Canadian regions sought similar public and private support for their strategic gateways and corridors. In 2007, the federal government responded to this countrywide interest with a *National Policy Framework for Strategic Gateways and Trade Corridors*. This Policy sets out as key criteria for support as being: "integration on several levels – across modes of transportation, between investment and policy, public and private sectors, and among levels of government" (Transport Canada 2007: 10). The *National Policy* provides a $2.1 billion infrastructure fund to support a limited number of strategic national gateways, key intermodal linkages and significant border crossings.

In 2004, the Halifax Gateway Council was established to tap into the potential Asian container trade coming through the Suez Canal to North America's East Coast. Subsequently, the Halifax Gateway Council was broadened into the Atlantic Gateway. A memorandum of understanding (MOU) outlining the collaboration between the federal government and the Atlantic Provinces in support of the Atlantic Gateway was signed in October 2007. Its objective is to develop an integrated multi-modal transportation system to support international trade through key gateway ports such as Halifax, Saint John, St. John's and Belledune.

A similar federal-provincial MOU establishing the Ontario-Quebec Continental Gateway and Trade Corridor was signed in July 2007. The Continental Gateway includes major St. Lawrence and Great Lakes ports. The Continental Gateway's focus is on integrating the region's strategic ports, airports, intermodal facilities, border crossings and essential road, rail and marine infrastructure with Canada's other two gateways and corridors. Montréal is positioning itself for significant growth in container throughput as part of the Continental Gateway strategy. The Montréal Port Authority's "Vision 2020" calls for a tripling of its annual container handling capacity to 4.5 million TEUs (20 foot equivalent units) (Swift 2009). The Port of Sept Îles in the Gulf of St. Lawrence has also examined the feasibility of developing a major container hub port serving the Continental Gateway as well as the Canadian and US East Coast via short sea shipping feeder vessels (Ircha 2005).

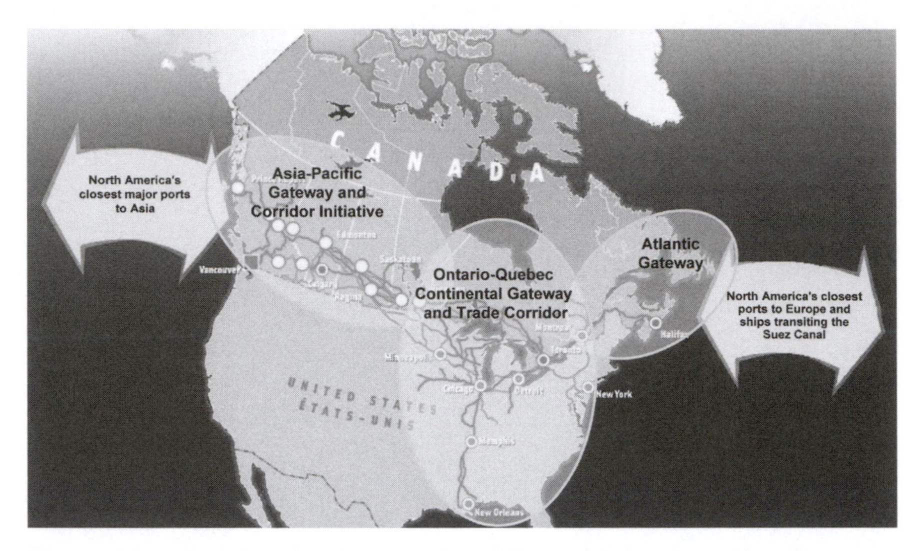

Figure 14.1 Overlapping Boundaries of Three Major Canadian Gateways (Transport Canada 2009)

Source: Canada's Gateways (www.canadasgateways.gc.ca).

When one considers the physical connectedness of the overlapping boundaries of Canada's three major gateways and trade corridors and their north-south connections, as shown in Figure 14.1, it appears that a consolidated national transportation strategy or plan could emerge.

Following the lead of the Vancouver and Halifax Gateway Councils, many sub-regions and municipalities within the trade corridors formed their own more localized gateway councils. In recent years, a plethora of such gateway councils have been established, including: Greater Vancouver (1994), Halifax (2004), Southern Ontario (2006), Manitoba International (2007), Southern New Brunswick (2008), Arctic (2008), Port Alberta, Edmonton (2008) and Sydney Ports (2008).

Several container hub port proposals have emerged in the Atlantic Gateway. For example, despite current surplus capacity at existing container terminals in Halifax and Saint John, Nova Scotia's Premier touted the benefits of developing a new private sector $300 million container terminal on a green field site at Melford in the Canso Strait area as part of the Atlantic Gateway initiative (Peters 2007). The Melford facility has received environmental approval to proceed (Canadian Press 2008) and expects to be operational by 2012 (Peters 2009a). Sydney and Searsport Maine soon joined the queue, seeking to develop new container terminals in their ports (Peters 2008a; Williams 2007).

These embryonic gateway port initiatives in the current Atlantic Gateway may not fit the federal government's key criteria of an integrated system. As a result there have been calls for a broader and more inclusive approach to Gateway development. As pointed out by Captain A. Soppitt, President and CEO of the

Saint John Port Authority: "we need cooperation and everybody pulling together to make it work. It's not about drawing container traffic out of Halifax or vice versa. If this growth is coming, we're going to need all those facilities, and how do we maximize all of those assets in the region?" (Horbie 2007: 16). The lack of integration was reflected in an Asia-Pacific Foundation study of the Atlantic Gateway:

> … the region does not see itself as global hub-and-spoke to the global economy. Each province views opportunities as event driven, driven by provincial concerns, usually with a short-term political advantage. The region rarely looks at global advantages, where cooperation and coordination are required (McMillan 2006: 9).

Trade corridors and gateways provide an essential means of encouraging enhanced integration, collaboration and co-operation in infrastructure investments by private and public partners. Private sector transportation operators, terminal managers and their clients are key to setting priorities for essential transportation infrastructure to overcome bottlenecks and expedite service. As the federal government pointed out:

> … gateway councils and other stakeholder-driven forums for consensus-building, planning and sound governance and accountability are also key to advancing regional strategies with national benefits … Actions should complement current market-oriented transportation policies, with government creating a positive climate for private investment in gateway infrastructure, while safeguarding the public interest (Transport Canada 2007: 12).

The integrative success of APGCI shows that cooperation, collaboration and coordination among many transportation stakeholders is achievable in the Canadian context.

4. Global – Local Concerns

Globalization has led to a paradoxical situation where the broader economic benefits of gateways and trade corridors often has come at the expense of local communities. As regional and national trade and business groups have increased their economic opportunities through growing international trade via Canada's (and North America's) gateways, local community support for these initiatives has often decreased (Hall 2007b). This reflects a process of "global change, local pain" (McCalla 1999b).

APGCI's success mirrors this global – local dilemma. Port Metro Vancouver officials often speak of the challenge they face in collaborating with the 16 Lower Mainland host municipalities surrounding their extensive port. Community concerns include environmental and community degradation from air, noise, and light pollution, roadway traffic congestion, railway crossing delays from lengthy double-stack container trains, lack of public access to waterfront lands, and First Nations land claims. Such concerns reflect how local communities suffer in allowing inland consumers to benefit from the international trade coming through the port. Port Metro Vancouver officials refer to the "social license" they received from local communities to allow their port to operate. Port officials are aware that they must take appropriate steps in their operations to mitigate their negative externalities to retain their "social license".

Prince Rupert's Fairview Container Terminal opened in 2007 and provides a good example of a purely "pass-through" port. In this case, containers are loaded directly onto double-stack railcars in an express service to the US Mid-west via Chicago. There is no expectation that containers will be destined for Prince Rupert and its immediate hinterland. Prince George and Edmonton, located on CN's route from Prince Rupert, are developing inland container ports. The expectation is that some of Prince Rupert's containers may be diverted to other North American locations as well as encouraging regional economic development by filling Asian-bound empty containers with local export goods (lumber, mineral resources and so forth). Currently, the major benefit to Prince Rupert from their growing container terminal seems limited to the port labor involved in container handling (Wilson and Summerville 2008).

Gateway councils can help address community concerns. The Vancouver Gateway Council identified community irritants and transportation bottlenecks and made them APGCI infrastructure investment priorities. In this manner, gateway councils provide an essential local forum for addressing transportation irritants and needs. For example, the Southern New Brunswick Gateway Council's aim is to "reach consensus on the high priority transportation and gateway needs for the entire region and approach government on this basis supported by other providers such as ports, airports and provincial governments" (Peters 2008b: 29).

5. Adding Value through Gateways and Corridors

Local gateway councils add considerable value in transportation corridors by identifying and prioritizing specific bottlenecks and pressure points in their multimodal transportation system. For example, Nova Scotia gateway councils in the Atlantic Gateway might suggest needed railway improvements to serve the proposed new container hub ports at Melford and/or Sydney as well as enhancing rail connections through Saint John and the New England states to better access the US. Similarly, the Southern New Brunswick Gateway Council might reinforce the need for crucial highway improvements from Saint John to St. Stephen on the Maine border to support enhanced US trade from Atlantic Canada. On a broader scale, gateway councils across the country could exert pressure on the federal government to address cross-border constraints on cargo through the "thickening" US border.

An Atlantic Gateway Council comprised of 13 prominent private sector members has been created. The Council represents some of the region's largest transportation firms and users who offer a strong business voice in the Atlantic Gateway complementing existing local gateway councils. The Atlantic Gateway Council sees its role as developing regional priorities to devise a list of key transportation infrastructure projects for federal and provincial consideration (Penty 2009).

A similar regional council, the St. Lawrence-Great Lakes Trade Gateway Leadership Council, has been established for the Continental Gateway (Frederick 2008). This Leadership Council has 74 business and community leaders from Ontario and Quebec. They recently completed a comprehensive study of the major transportation needs in their corridor (IBI 2008). The Southern Ontario Gateway Council also published an extensive study of their local transportation needs to devise a list of priority projects (Taylor and Hackston 2008). These major transportation studies outline the key transportation priorities to guide public and private investment.

A key maritime issue for both the Continental and Atlantic Gateways is the development of short-sea shipping through the St. Lawrence Seaway and on the Great Lakes. There are many impediments to developing this maritime transport service including: the closure of the Seaway during the winter months, Canada's 25 percent duty applied to foreign flag vessels acquired for domestic use, navigation fees, pilotage charges, the US Harbor Maintenance Tax (HMT) applied to landed goods in US ports, and significant cabotage restrictions on cross-border trade (Frost and Roy 2008; Roy and Frost 2009).

The winter closure of the St. Lawrence Seaway forces short-sea shippers to find alternative short-term transport on competing modes (such as road or rail) during this seasonal period. Such transport alternatives do not come cheap – typically competing modes will not offer volume or other discounts for such discretionary traffic. Short sea shipping advocates argue that the 25 percent duty on foreign built ships along with pilotage and navigation charges all add considerable costs, which tend to stifle these newly emerging endeavours. The St. Lawrence Seaway is promoting short-sea shipping by reducing its tolls for these services (Peters 2009b).

The US Harbor Maintenance Tax (HMT) causes concerns for short-sea shipping on both sides of the border. HMT is based on 0.125 percent of cargo value landed at a port. Its purpose is to raise funds for harbor dredging; however, since short-sea shipping inevitably entails shallow draft vessels, no benefit is received by these coastal ships from the HMT. In addition, seeking to avoid HMT has led cross-border shippers to use trucks rather than marine ferry alternatives. Several unsuccessful attempts have been made in the US Congress to reduce HMTs effect on short-sea shipping (Blenkey 2009).

A more significant impediment to cross-border short-sea shipping is the cabotage restrictions imposed by both Canada and the US. NAFTA discussions failed to eliminate or relax marine cabotage provisions. Under NAFTA, Canada and Mexico did agree to liberalize their cabotage rules with each other, but the US chose to remain outside this arrangement. Current regulatory restrictions hamper the development of a short sea-shipping regime on the East and West Coasts and the Great Lakes. It is unlikely that an international operator can viably mount a one port per country shuttle service to effectively compete with overland intermodal services. Given the lengthy history of unsuccessful attempts to amend the US Jones Act to liberalize cabotage in coastal shipping, it appears that such dramatic regulatory change will not occur in the near future. However, in April 2006 an international Marine Conference in Vancouver addressed the feasibility of using short-sea shipping to alleviate chronic rail and road congestion in North American ports. It is possible that institutional changes emerging from this conference could generate further support for short-sea shipping. As an example, a tri-partite declaration of the three NAFTA nations arose from the conference to "provide a general framework for the establishment of the North American Short Sea Shipping Steering Committee to … ultimately develop a North American short sea shipping strategy" (Declaration 2006).

Despite these impediments, new short-sea shipping initiatives are being planned or are underway. Recently, Great Lakes Feeder Lines acquired a German built multipurpose vessel to move containers and other cargo between Halifax, Montreal and Toronto. Having paid the 25 percent duty to convert the ship to the Canadian flag, it can now load freight in US ports but can only drop it off at a non-US port, due to the Jones Act cabotage restrictions. Despite difficulties in attracting cargo from existing road and rail services, the shipping line plans on expanding its fleet to five vessels over the next three years (Gillis 2008). The expectation of developing short-sea shipping opportunities on the Great Lakes led the small US port of Oswego New York to seek funding for a container handling facility. Oswego's goal is to tap into short-sea feeder services from Montreal and Halifax (Edmonson 2009).

6. An Evolving National Transportation Plan

Many gateway councils are focusing on local transportation priorities. Regional trade corridor councils, such as APGCI, Atlantic Gateway Council and Continental Leadership Council, comprised of private sector transportation providers and users along with representatives from local gateway councils and other stakeholders, could blend local transportation priorities into a comprehensive set of regional priorities for federal and provincial funding consideration. In turn, representatives from the regional trade corridor councils could come together to form a new National Trade Corridor and Gateway Council. This National Council could help set national transportation priorities as well as provide direction on inter-corridor and international transportation issues.

On a countrywide scale, a National Trade Corridor and Gateway Council could consider provincial and federal transportation regulations, provincial fiscal regimes, and provide guidance by setting priorities for significant transportation infrastructure investments and regulatory reform. Dealing with inter-corridor issues for example, could lead the National Council to encourage the federal and Quebec governments to fund the twinning of Highway 185 from the New Brunswick border to Rivière-du-Loup. This critical link lies between the Continental and Atlantic Gateways. It is the only un-twinned portion of the Trans-Canada Highway between Halifax and Central Canada (Frost 2008).

A National Trade Corridor and Gateway Council could assist the federal government in determining continental and international strategic transportation investments to support Canada's international trade. For example, trade between Atlantic Canada and the US has flourished with most exports to the US traveling by truck (partly due to poor rail and short-sea shipping connections). Maritime trucking firms often use a "triangular" routing moving Atlantic Canada products to the US, followed by exports from the US to Ontario and Quebec and Central Canadian products being brought back to the Atlantic region. For well over a decade, there have been calls for the development of a trans-Maine highway linking the Maritimes to northern New York State and Central Canada. Such a new highway would provide a speedier route for transporting Atlantic Canada products to markets in the US and Canada. The US has taken steps to designate this potential highway as a high priority east-west route as well as providing funding for specific studies, including a multi-modal trade corridor from Halifax to northern New York and the Ontario border (New Hampshire 2008). Given the importance of this proposed multi-modal corridor to the Atlantic Gateway, would it not make sense for the federal government to assist the US in developing this route? A National Trade Corridor and Gateway Council could encourage the federal government to support this and other similar Canada – US integrated trade development initiatives.

On a broader international scale, the Atlantic Gateway expects considerable trade growth from India and other Southeast Asian countries through the Suez Canal to Halifax and other regional ports. A critical factor stifling India's trade development is the deplorable state of that country's multi-modal transportation systems. Canadian firms are internationally recognized for their transportation expertise. Would it not make sense for the federal government to assist our international trading partners in India and South-East Asia by offering expert support to address their inland transportation needs? (Leroux 2007). Again, a National Trade Corridor and Gateway Council could assist the federal government in identifying strategic transportation aid and development projects, which in turn supports Canada's economic development.

A National Council could also assist in supporting US efforts to harmonize their differing regulations and taxation regimes to derive a set of NAFTA standards in partnership with Canada and Mexico. Such standards would enhance productivity and efficiency for multi-modal carriers across the continent.

By tackling these national and international initiatives, a National Council would support the federal government in achieving its stated goal: "as gateway strategies mature, the synergies and linkages among them will also be developed, further deepening their contribution to the national pursuit of economic competitiveness" (Transport Canada 2007: 14).

7. Conclusion

This paper suggests that a centrist, top-down development of a national transportation strategy or plan is a non-starter. However, the federal government's support of designated trade corridors and gateways, which involve public-private partnerships, may ultimately lead to such a national plan. By encouraging the development of a hierarchical arrangement of gateway councils from local gateway councils to regional trade corridor councils to a National Trade Corridor and Gateway Council, the federal government could create the means for deriving, from the bottom-up, a national (and possibly a continental) strategic transportation plan. Care would have to be taken to ensure that other broader transportation concerns beyond the scope of local, regional and the National Council be considered in devising a national plan.

A national strategic transportation plan could assist governments at all levels in meeting the many challenges of the new global economic context as well as ensuring collaboration and coordination in domestic, national and continental transportation interests. Now is the time to create a national plan in anticipation of a return to a more thoughtful and longer looking policy-making environment in our new global economic context.

Time will tell whether the current evolution of gateway and trade corridor councils will result in a comprehensive national strategic transportation plan. Such a plan prioritizes key infrastructure and service improvements and addresses broader national and continental issues to add value and enhance Canada's domestic and international trade. A further step would be the creation of a NAFTA Trade Corridor and Gateway Council to focus on transportation issues and priorities on continental basis. Such national, and indeed, international councils comprised of transportation providers, users, government agencies and other stakeholders are essential for prioritizing transportation infrastructure and service needs in the light of the massive infrastructure stimulus packages being devised to meet the challenges of the new global economy.

8. Acknowledgements

The Social Sciences and Humanities Research Council of Canada's Major Collaborative Research Initiative program on "Multilevel Governance and Public Policy in Canadian Municipalities" provides ongoing support for the author's research on federal ports and waterfront lands.

Chapter 15
Hinterlands, Port Regionalisation and Extended Gateways: The Case of Belgium and Northern France

Jacques J. Charlier

1. Introduction: The Revival and Growing Complexity of the Hinterland Concept

Extended gateways are seen by the Flanders Institute for Logistics as one of the solutions to alleviate congestion at major European container ports, especially Antwerp. In short, the idea is to develop, at a series of key locations along the trade corridors between these seaports and their hinterlands, a few logistical centres to and from which containers will be shuttled in large volumes by barge and/or rail transport. These extended gateways are the product of port regionalisation and they should be considered as functional satellites of the said seaports, wherein the latter could even invest directly or indirectly in order to strengthen the links with these technical and logistical bridgeheads in their continental hinterlands. The concept was first applied to Flanders, and then extended to Wallonia (in Northern and Southern Belgium, respectively) associated with the ports of Antwerp and Zeebrugge. The concept has also been adopted recently by the port of Rotterdam in respect to its Dutch and German hinterland.

In France the adoption of the concept has been slow with the first major example of an extended gateway being in the Nord-Pas de Calais region serving primarily the Benelux seaports of Rotterdam, Antwerp and Zeebrugge. The extended gateway is organised around the river port of Lille and the nearby trimodal logistical platform of Dourges. These two facilities also serve the French seaports of Dunkirk and Le Havre, but they are often seen as Trojan horses on French ground for the Benelux seaports. However the main French seaports, especially Le Havre, are now adapting this imported concept in order to try to catch up some of the ground they lost in the last decades. On the one hand, Lille and Dourges are strengthening their dominant position in the Île-de-France region for Le Havre by organising high capacity river shuttle services to and from the Gennevilliers port, where the largest intermodal and logistical platform serving Paris and its region is located. On the other hand, similar functional satellites are planned along the Seine-North canal (in between Paris and Dourges), which recently received a final green light. It remains to be seen whether Le Havre and Dunkirk will benefit more than the Benelux seaports from these additional facilities that will bridge the gap between the Lille and Paris regions.

The emergence of these extended gateways resonates with an ongoing clarification and elaboration of the hinterland concept in the literature on ports and shipping. Dramatic changes took place during the transition of deep sea trades from conventional general cargo to containers (Vigarié 1999), and containerisation is now the rule on the East-West and on many North-South trades (Culady 2006; Levinson 2006; Frémont 2007b; Rodrigue and Notteboom 2009b). In the early days, the literature mainly dealt with the geography of the container seaborne trades and with the ports that were served by container ships. But little was said about what was happening in the hinterlands of these ports, whose softening or disappearing was even predicted (Blumenhagen 1981; Hayuth 1982; Foggin and Dacey 1984; Hayuth 1987; Slack 1989c). Seaport hinterlands were first studied in the days of conventional general cargo (Morgan 1951; Patton 1958; Weigand 1958), and this is perhaps why they appeared as a concept of the past.

However hinterlands were gradually re-introduced when the research community started to understand that the battles were not only fought at sea and in ports, but also on land (Charlier 1992; Slack 1998; Notteboom 2000; Vigarié 2004; Debrie and Guerrero 2008; Rodrigue and Browne 2008; Notteboom and Rodrigue 2009). More refined concepts were introduced recently, namely port sub-harborisation (Notteboom and Rodrigue 2004), port regionalisation (Notteboom and Rodrigue 2005; Notteboom 2006c) and, as the tool of this regionalisation, extended gateways (Van Bredam and Vannieuwenhuise 2006). These are intermodal and logistical bridgeheads of seaports at the heart of their (container) hinterlands, and they are designed to relieve physical congestion at the main container ports. This function was already identified by Slack as early as 1999 when he was referring to (inland) satellite terminals, but what makes extended gateways more complex is that they combine intermodal and logistical facilities.

The logistical dimension was taken into consideration by port and maritime geographers when shipping lines extended their services from *port to port*, to *door to door* and finally to *floor to floor* (Notteboom and Rodrigue 2007; Frémont 2008). The massification of hinterland flows along inland corridors (be it by container barges or by unit train services) also received a great deal of attention (Notteboom 2002b; Gouvernal and Daydou 2005; Franc 2007; Notteboom 2008), as well as the intermodal facilities in the hinterlands. These are either pure transfer facilities or are associated with logistical services, and a lot of terms have been used in this respect, from inland container terminals (Hayuth 1980; Rodrigue and Notteboom 2009c) to intermodal nodes (Ridolfi 1986; Charlier and Ridolfi 1994), inland load centers (Slack 1990, 1994) and dry ports (Debrie 2004; Rosso *et al.* 2009). This is perhaps a false dichotomy, as the massified transport of containers along transport corridors in the hinterland and the intermodal nodes structuring the latter are two interrelated faces of the same coin (Charlier and Ridolfi 1994; Notteboom 2008).

2. Extended Gateways: A Belgian Concept

The port of Antwerp is, by far, the largest logistical complex of Europe (Durveaux 2004), with currently more than 5.5 million square meters of covered storage space. It is also among the largest container ports of the world, where the share of boxes stuffed or unstuffed locally is the highest (about 20 per cent of the "continental" containers, excluding transhipment and feedering). Moreover, Antwerp is more than a major maritime container port with more than 8.5 million TEU in 2008; it is also the largest dry port in the world. Its logistical facilities also handle a large volume of marine containers loaded or unloaded in Rotterdam or Zeebrugge and shuttled between these two ports and Antwerp by barge, train or truck. Our own educated guess for this indirect traffic, based upon industry sources, is around 750,000 TEU and 250,000 TEU respectively.

2.1 A Concept Born in Flanders

In search of cheaper land and manpower, some logistical facilities directly or indirectly linked to maritime imports or exports have been built in the recent years outside the port of Antwerp, either in its vicinity or deeper in its hinterland. These are often linked to intermodal terminals that also serve more generally the regional hinterland (Belgium is a federal state with three political regions: Flanders, Brussels and Wallonia). There are several gateways in Flanders: four seaports (Antwerp, Ghent, Zeebrugge and Ostend) and two freight airports (the national airport of Brussels/Zaventem and the regional airport of Ostend). These are linked to their regional and extraregional hinterland by a very dense network of waterways, railways and motorways, along which a series of the above-mentioned inland intermodal terminals can be found (Macharis 2002). Moreover, a series of geographic clusters can be identified in the field of logistics along these major corridors or in their vicinity.

In a conceptual vision of "Flanders Logistics" (paralleling "Nederland Distributieland" around the Dutch seaports and airports), the Antwerp-based Flanders Institute for Logistics (Vlaams Instituut voor de Logistiek or VIL) has identified three levels of densities for these geographically clustered logistical facilities: first order (primary hot spot), second order (secondary hot spot) and emerging (runner up). They extend primarily in the Northern part of Flanders, in between the coastal region and the Campine (Kempen) region, and also near Brussels (a separate city-region, land-locked within Flanders). However there has been a growing interest in recent years for locations in the Southern part of Flanders in search for cheaper land and labour, as well as for less congestion (and in some cases for better access to/from customers outside Flanders).

The name "extended gateways" has been used in Flanders for some time in regards to this conceptual approach (Van Breedam and Vannieuwenhuyse 2006). It is now widely quoted inside and outside academic circles, including in the Netherlands where it has been adopted recently in respect with the port of Rotterdam and its hinterland (Konings *et al.* 2009). However, the current Flemish or Dutch extended gateways are not the product of a planned policy of sub-harborisation, as the extended gateway concept has been applied there *ex-post* to an existing situation. The same can be said about Paris and Le Havre in section 3.2, but there are three cases where new intermodal and logistical developments have been or will be taking place *ex-ante* around the extended gateway concept: the Delta 3 platform at Dourges near Lille in Northern France (discussed in section 3.1), those planned further South along the inland canal that will connect this region with the Seine basin (section 3.3), and the Trilogiport platform near Liège, in Eastern Wallonia (section 2.2). The last example should be considered as the first real Belgian planned application of the concept. Other smaller similar facilities are being built or are planned elsewhere in Wallonia, including near La Louvière.

2.2 Wallonia in the Era of the Extended Gateways

Liège is Belgium's fourth largest city, and it has recently become a major hub for air freight. With 16 million tons handled by inland barges in 2008, its port is among the top three inland ports in Europe, along with Duisburg and Paris. Several intermodal terminals already serve it, but they are small and their location is not optimal. On the one hand, there is a local need for better intermodal facilities, better located for barge traffic on the Albert canal from/to Antwerp (or through this seaport and then via the Scheldt-Rhine canal from/to the port of Rotterdam, which is also directly accessible from Liège via the Juliana canal). On the other hand, the Walloon regional government is pushing logistics as one of the five major avenues for its economic redevelopment in the context of its current economic redevelopment plan. The first steps in this direction have been crowned with success, as Wallonia has been identified as one of the three best European locations for logistical facilities, on equal footing with Flanders and Northern France (Rossall and Barrie 2008). Hence the recent decision was made to develop a 110-hectare platform aptly named Trilogiport, combining a 15-hectare trimodal terminal along the above-mentioned Albert canal and an adjacent dock (with an overall quay length of 2,100 m) and a significant amount of logistical space (40 ha) immediately behind it to attract European Distribution Centers). Work is in progress and the first intermodal and logistical facilities are due to be completed by 2013.

What is unique in the Trilogiport project is that it will be developed jointly between the port authorities of Antwerp and Liège, together with a local development agency (SPI+). These partners established an Economic Interest Grouping (EIG) in 2006 with "the aim to guarantee a seamless system of distribution downstream from Antwerp's cargo unloading system" (http://www.port-autonome.be/fr/pages/trilogiport.aspx). On this web site, the advantages for the two parties are shown as a win-win situation: "this EIG is primarily concerned with allowing Port of Antwerp's customers who do not need a coastal wharf to become established in its natural hinterland and with improving inward and outward mobility in Antwerp. The EIG will also enable the Liège Port Authority to enjoy the benefits of major international companies that are drawn by the Port of Antwerp status and decide to set-up inland, thereby enhancing the Liège Port Authority's international profile". Transnational companies and world-class logistical services providers are clearly targeted, and the Port of Liège Authority has already started its marketing efforts by attending the MIPIM real estate fair in Cannes (the largest in Europe) in March 2009 and the Transport Logistics fair in Munich (one of the largest in the world) in May 2009. In the latter case, the Trilogiport project was even featured on two booths – the one of the Walloon Region and the one of the Port of Antwerp – showing thus the highest interest in extending the spatial coverage of its network of extended gateways outside Flanders.

Firm contracts have been signed recently with major players: on the intermodal front with a consortium established by DP World, Manuport (an Antwerp-based handling company) and Water Container Transport (the operator of a barge/road container terminal on the Albert Canal); and on the logistics front with Deutsche Lagerhaus Trilogiport for 30 ha (the Belgian branch of a German developer, that will lease its sheds to international 3PL providers) and with Warehouses De Pauw for another 10 ha (a Belgian developer whose customers will be attracted on a national or regional base). The success has been such that the port of Rotterdam has recently asked to be associated with Trilogiport or, more realistically, with its planned extension (Trilogiport 2). This was due to take place next door along the Albert canal on the site of a steel plant that was due to close, but its closure has been postponed and a new location is thus required.

There are many examples of informal partnerships between seaports and hinterland nodes, but for Antwerp, this is the first example of a formal strategic partnership with an inland intermodal and logistical centre bridgehead in its hinterland. Other examples in Northern France are considered in the next sections. It should be noted that another platform of this type (Garocentre South) is currently being developed in the Western part of Wallonia, near La Louvière, and that the elements are the same. These include a trimodal terminal for containers and steel products, as well as a 50-hectare "Magna Park" marketed internationally by the British logistical developer Gazeley. This will fit well in the Belgian network of extended gateways, as it is located 40 km to the South of Brussels, in between Liège and Lille in Northern France.

3. At the Crossroads of Seaport Hinterlands: Northern France

France is a large country, whose inland regions are served from a series of national and international port gateways located on three seaport ranges: the Mediterranean range (with Marseilles, the largest French port), the Atlantic range (with several mid-sized seaports in between Bordeaux and Cherbourg), and the Western part of the Northern range (with a series of French and foreign seaports in between Le

Havre and Rotterdam, including Dunkirk, Zeebrugge and Antwerp) (Debrie and Guerrero 2008).

In the second part of this chapter, we will show that several extended gateways, some of the *de facto* type and other of the planned type, can be found in the Northern part of the country in between the Belgian border and Paris. Again, waterways are or will be playing here a major role in the relationships between the seaports and this part of their respective hinterlands.

3.1 The Nord–Pas-de-Calais Region: From Lille to Dourges

The French Nord–Pas-de-Calais region cannot be classified as being within the exclusive hinterland of any single port, including its "natural" regional gateways of Dunkirk and Calais (Charlier 1983). On the contrary, it is the region where hinterland competition is the fiercest with the Belgian ports on the one hand (Antwerp, but also Ghent and Zeebrugge) and the Lower Seine seaports on the other hand (Le Havre and, to a lesser extent, Rouen). Historically, the Lille-Roubaix-Tourcoing conurbation has been the intermodal and logistical node of the Nord-Pas-de-Calais region. The historical port of Lille is still a major player, thanks to its good river connections with the Belgian seaports and the Lille Container Terminal that was one of the very first to be developed at a French inland port (Joan 2008). However, the newest developments in these fields have been taking place somewhat outside the conurbation, where ample space was available and where the proximity of the city was not an issue in respect with traffic congestion and environmental problems.

Liège's Trilogiport project is very much inspired by the 300-hectare Delta 3 greenfield project whose first phase became operational in 2003 in Dourges, to the South of Lille. It is located at a unique crossroad of a large canal, a major railway line and a major motorway (Figure 15.1). First class transport connections are offered in all directions (including by road to/from the UK). For containers, Dunkirk, Zeebrugge, Antwerp and even Rotterdam can conveniently be served by barges, albeit of a more limited size than in the Rhine-Meuse-Scheldt system. This is why this facility has also been planned as a trimodal one, with a current quay length of just 250 m which could easily be extended when needed especially when the planned Seine-Nord canal will be in service. Associated with a marshalling yard featuring 15 tracks, its rail/road terminal is impressive, with seven 750 m-long tracks served by four portal cranes, supplemented by a trimodal gantry crane at the river berth where intermodal trains are also loaded/unloaded. Rail services (some for marine containers, but mainly for continental intermodal units, namely continental containers or swapbodies) are offered from/to a number of French cities (including Le Havre and Marseilles) as well as from/to three international origins/destinations. These include one in Germany, one in Italy and one in Spain; in the future, the UK could also be served by direct trains running through the Channel tunnel.

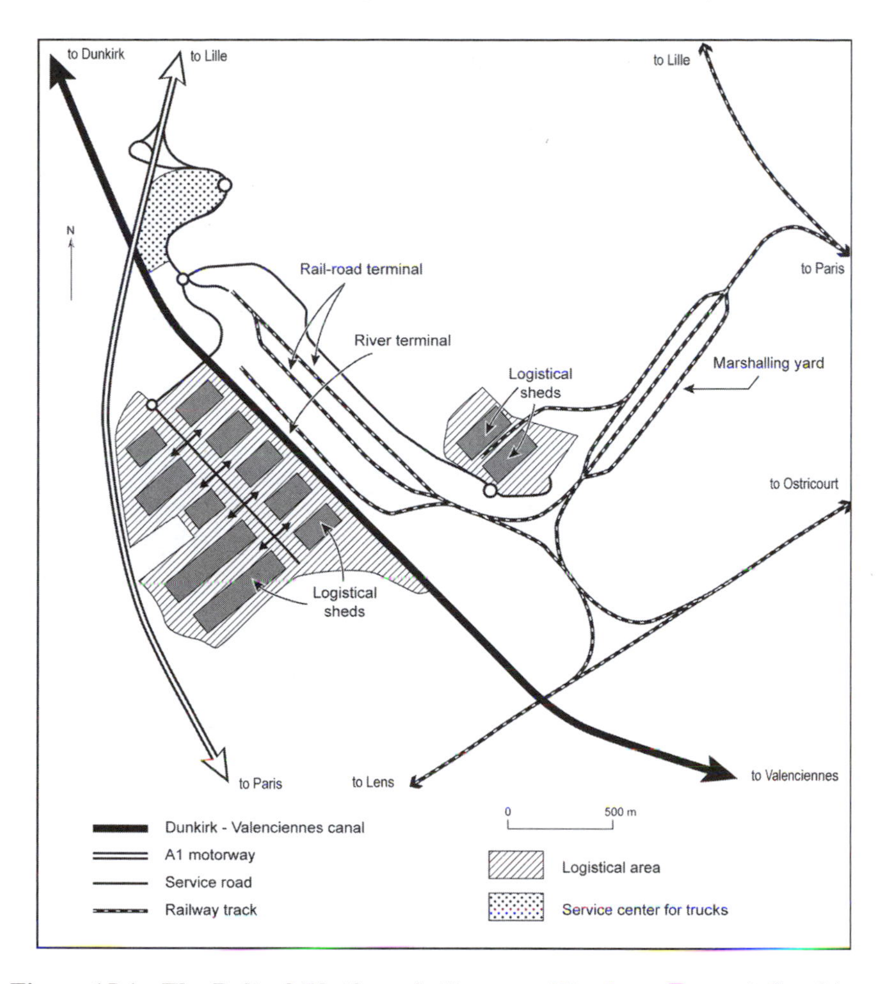

Figure 15.1 The Delta 3 Platform in Dourges (Northern France) Combines Trimodal Terminal and Logistical Facilities

Because of this mix of maritime and continental intermodal units, Delta 3 and the nearby port of Lille do not form a "pure" extended gateway, serving only a series of French, Belgian and Dutch seaports competing for the traffic of marine containers. This suggests that port economists and geographers should stop considering hinterlands as closed systems oriented exclusively towards seaports: on the one hand, in terms of port regionalisation, extended gateways are contributing to strengthen the "grip" of one, two or several seaport(s) in its (their) exclusive or shared hinterland(s); but on the other hand, in Europe as in North America (where intermodal traffic has also a strong position on the domestic transportation market), many intermodal terminals or clusters of such terminals are dual purpose facilities, and serve at the same time marine container traffic and continental intermodal flows whose geographies are quite different.

On the logistics front the Delta 3 platform is equality impressive, as major logistical developments have taken place around its intermodal terminal. There are currently 330,000 m^2 for sheds arranged into three separate zones; a small one (Distrirail Delta 3) is served by road, but also by rail for conventional traffic, and two adjacent, larger ones are served only by road directly from the adjacent motorway (Distripôle Delta 3 and Zone L.A. Nord). The success has been so great that a 115-hectare extension, with another 220,000 m^2 for sheds, is already being considered (www.delta-3.com).

As in the case of Wallonia, nothing would have been possible without the strong support of the regional public authorities. First they joined their forces in a "Comité syndical" whose members are the Nord-Pas-de-Calais Region and its two Departments, the Lille Metropolitan Urban Community and four local municipal communities. Thereafter, in order to build the platform, they established in 2000 a public-private "Syndicat mixte", SAEM Delta 3 (where SAEM stands for "Société Anonyme d'Economie Mixte"). Its other members are three financial institutions (the national Caisse des Dépôts et Consignations, representing the French State, and two regional saving banks), the regional and local Chambers of Commerce and Industry, and finally a subsidiary of the French national railway company (SNCF Participations). The SAEM Delta 3 is not the operator of the trimodal terminal; rather it is managed through a long term lease by LDCT (Lille Dourges Container Terminal), a consortium of intermodal service providers associated with NCS. NCS is an EIG established by the ports of Dunkirk and Lille to operate barge services. The strategic link between Dunkirk and Dourges is thus indirect and rather limited in scope. Nor is Delta 3 a public logistics services provider; the above-mentioned Distrirail Delta 3 and Distripôle Delta 3 logistical zones were built and marketed jointly with two private developers, whereas private companies have invested directly in their own logistical facilities in the L-A Nord zone (and will probably do so in the planned extension).

3.2 An Old Couple: Le Havre and Paris

The Lower Seine seaports of Rouen, followed by Le Havre, have been for centuries *the* oceanic gateways for Paris and the Ile-de-France region (Charlier 1991; Debrie and Guerrero 2008). In the same way, as for the Rhône axis in between Marseilles and Lyons, there is a strong geographical determinism, combining the rather short distances between the two main French ports and their two respective bridgeheads in the French hinterland, and a trimodal transportation corridor (even quadrimodal, when taking also the pipelines into consideration) along the valleys linking these two pairs of a seaport and an inland city. Unfortunately, the Seine and Rhône rivers are not (yet) connected with Northern and Eastern France by waterways allowing the transit of large gauge inland craft (Frémont *et al.* 2008). Thus, Paris as well as Lyons are like cul-de-sacs for river navigation. Worse, in both cases, the two cities are an obstacle for either the West-East or South-North transit of goods by rail or road because of the lack of capacity and congestion of their respective rail and road networks.

The largest of the public facilities provided by the Port Autonome de Paris (PAP) is located at Gennevilliers, in the Western suburbs of the French metropolis. The capacity of the Gennevilliers intermodal terminal operated by Paris Terminal (a subsidiary of PAP) is currently being extended from 250,000 TEU/year to 450,000. The quay length for barge handling is being increased from 500 m to 900 m and the number of trimodal container gantries from two to three; at the same time, the length of the trackage available to load or unload intermodal trains (for marine containers, but also for continental units including swapbodies) is being extended from 1,600 m to 3,500 m. As pointed out recently by Frémont (2009b), such a terminal is a powerful tool for territorial integration in the Paris region. It should be noted that the Gennevilliers terminal is not the sole intermodal hub to the region. In order to offer a more comprehensive spatial coverage for this huge urban complex, two smaller intermodal terminals have been developed more recently at Bonneuil-sur-Marne, on the Eastern side of Paris. There is also a fourth terminal in Nogent-sur-Seine, further upstream but outside the territory managed by the PAP (because it is located in the Champagne-Ardennes region). Another three such intermodal facilities, where limited intermodal operations have already begun in some cases, are being currently developed by the PAP within the Ile-de-France at Limay, Bruyères-sur-Oise and Montereau, whereas another four are being considered at Archères, Evry, Saint-Ouen and Longueil-Ste-Marie.

These intermodal terminals are only one aspect of PAP's strategy to be a major player. The second and equally important element is that logistics facilities are being associated with these new terminal developments as much as possible. When taking all these intermodal and logistics platforms into consideration, Paris and the Ile-de-France as a whole appear to be a first-level extended gateway, on equal footing with Duisburg at the European level. At present, there is a shortage of suitable space, as land is very expensive around Paris, especially in "wet" locations, both of which limit PAP's expansion possibilities. Since new logistical facilities will soon be available along the Seine-Nord canal (see below) located in between Paris and Lille/Dourges, they will form an intermediate cluster of platforms.

3.3 Will Another Extended Gateway Emerge between Paris and Lille?

In April 2009, the project of the new Seine-Nord canal, designed for large dimension inland barges of the European Vb class and to be built under a PPP (Public-Private Partnership) system, got the final green light after years of studies and debates and despite a predicted modest return in strict economic terms. This is because of the "green" dimension of the project, whose economic value is hard to internalize. The decision to go ahead was also taken into consideration in the context of the "Grenelle Environment Forum" promoted by French public authorities (Joignaux and Courtois 2009). The proof that the concept of the "greening of transport" is more than an empty claim in today's European political and socio-economical context can be found in the fact that this green light was given by a right-wing government. It is interesting to note that a former left-wing government, including a Minister of the Environment from the French Green Party, had turned down (under the pressure of the greens themselves) the project of a similar canal linking the Rhine river with the Saône river, a subsidiary of the Rhône in 1997, thus denying convenient access to the Rhine basin for the port of Marseilles.

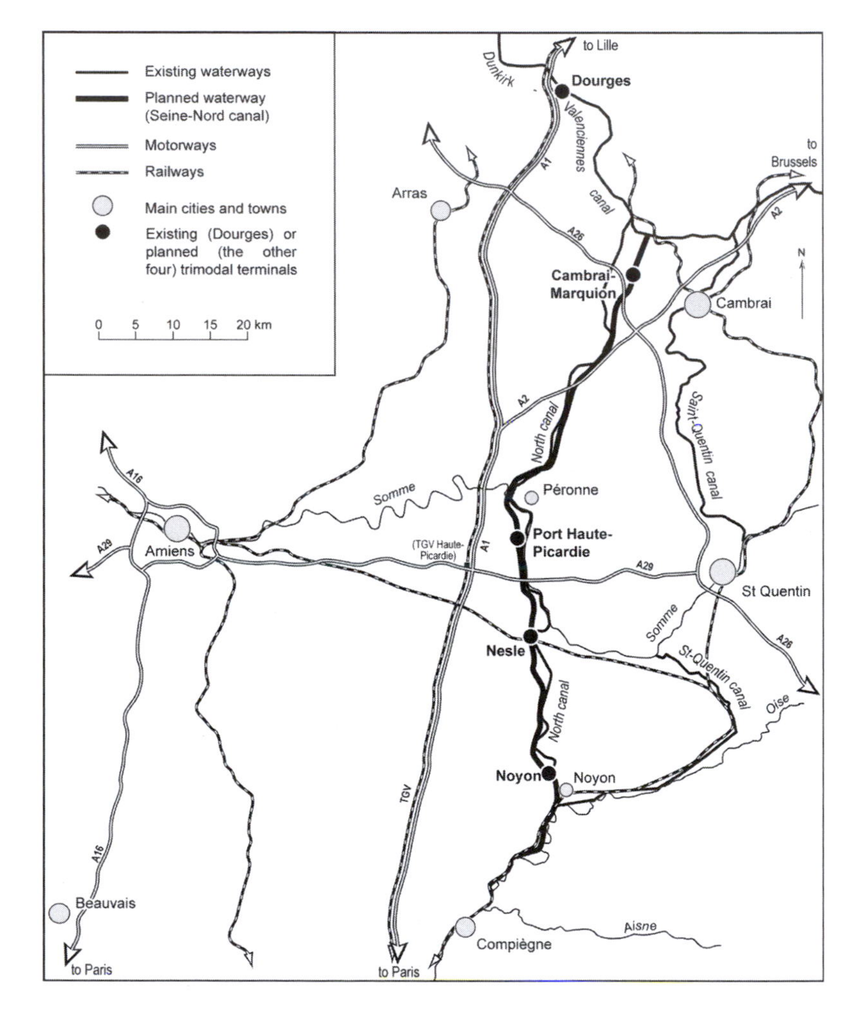

Figure 15.2 The Planned Seine-Nord Canal and its Intermodal/Logistical Platforms

In short, this new waterway (to be built, to be fair, in a less environmentally-sensitive region and in much flatter land than the now abandoned Rhine-Rhône project) will link Compiègne (on the Oise river to the North of Paris) to Aubencheul-au-Bac (to the West of Cambrai), a few kilometres away from the above-mentioned Dourges/Delta 3 platform on the Dunkirk-Valenciennes canal (Figure 15.2). The length of this missing link in the European network (replacing an existing more or less parallel smaller canal) is just 106 km, and its total land take will total 2,450 ha (including 360 ha for the four intermodal platforms mentioned below). It will feature seven locks and it will carry a wide range of goods, not just containers, as this new waterway will run through a rich agricultural region; as well, there is a big market for the North-South transit of dry bulk and conventional cargo. The current forecast for the overall traffic in 2020 is between 13 and 15 million tons and between 20 and 28 million tons in 2050.

The new waterway is part of a more comprehensive project (Seine-Scheldt) agreed between the French and Belgian authorities through, respectively, Voies Navigables de France (VNF), as well as the Flemish and Walloon regions (Bour 2008). At its Southern end, the Oise will be modernised to Seine navigational standards as far as Compiègne, and at the Northern end a series of similar improvements will take place in order to connect at the same standards with the ports of Dunkirk and Ghent (and from there either to Rotterdam and Antwerp or to Zeebrugge, in the latter case via an new canal to be funded separately by the Flemish government). From the region of Lille, access will also be improved to Wallonia and its new intermodal/logistical facilities.

In terms of regional development, this new French waterway should have positive economic effects for the region through which it will pass. A series of local port facilities are planned (including five for grain traffic), together with four larger, more or less evenly spaced, intermodal platforms. From North to South, these should be built near Cambrai, Péronne, Nesle and Noyon, and they will be served not just by the canal but also by rail. However, for the time being, VNF is not planning to invest only in logistics. One of the four platforms (Nesle) is intended to serve primarily industrial plants, with just a secondary logistical function. Two others (Cambrai and Noyon) should accommodate a balanced mix of industrial and logistical facilities, whereas only the fourth one (Péronne) is intended to be primarily oriented towards logistics. Located more or less at equal distance from the two ends of the canal, it will have a higher level of regional influence in the Picardie region, as the commercial name tentatively selected for this platform indicates: Port Haute-Picardie.

4. Conclusions

A refreshing wind of change is blowing in port geography, with a young guard of researchers exploring the most complex side of the port triptych, namely the hinterland. The old guard (of which this author is a member!) did not fully appreciate the many dimensions to be taken into account in its study until a paper by Charlier and Ridolfi (1994), later confirmed by Vigarié (2004). The numerous references associated with this chapter, most of which were written less than 15 years ago, show the extraordinary vitality of the research undertaken recently by maritime and port geographers and economists in this field. This is partly because they have realised that the container lines have become true intermodal players whose maritime and continental operations should now be considered on equal footing. The rise of logistics has made the situation even more complex, encouraging the said younger researchers to explore further all the upstream and downstream segments of the chain from producer to consumer. Hence arose the emergence of new concepts associated with the traditional hinterland concept, the richest of which is port regionalisation on the abstract level.

As shown in the chapter this concept should be associated, on the physical level, with another one, namely that of extended gateways. The case study offered here has shown how, after being proposed in order to structure *ex post* the territory of Flanders in the immediate hinterland of the port of Antwerp, the extended gateway concept is currently expanding *ex ante* into nearby Wallonia, then outside the Belgian borders, in between Lille to Paris (from Dourges to Port Haute-Picardie). However, the longer the distance from Antwerp is, the less exclusive is or will be the relationship with Antwerp, and the higher the competition is or will be with other seaports. Competition is coming from Rotterdam in the case of Liège and Dunkirk as well as Le Havre in the case of the French extended gateways. The Lille and Paris regions are now well established extended gateways, and impressive additional developments are taking place or are planned there in the intermodal and logistical fields. These will not remain isolated geographically, as another extended gateway will probably rise along the new Seine-Nord canal, bridging the current intermodal and logistical gap between Paris and Lille. This is probably where the level of interport competition will be the highest among seaports following independent strategies of port regionalisation. In turn, such a development will offer the first example of a shared extended gateway, whereas the earlier examples implied a more exclusive port-extended gateway relationship.

Chapter 16
Entrepreneurial Region and Gateway-Making in China: A Case Study of Guangxi

James J. Wang

1. Introduction and Literature Review

Like many common words, "gateway" has often been loosely defined and used. One of the more carefully defined concepts of gateway comes from Burghardt (1971), who considered a gateway as an entrance into or out of some area geographically, with a fertile hinterland on the one side and an infertile region on the other. The concept of a gateway has regained its popularity recently in the formulation of regional or even national development strategies in various countries, including the USA and Canada. From a transport geographic point of view, such a trend can be seen as the recognition and realization of geographical intermediacy. From a global supply chain (GSC) perspective, a gateway is regarded as a node in a GSC when the transport searches for the easiest, shortest, and lowest cost route (Gillen *et al.* 2007). Attracting GSCs to one or a set of gateways and corridors can also be interpreted as the state's efforts to reshape its accessibility pattern in order to gain from the "transit" economy.

A similar discussion can be found in world city literature, wherein a gateway city is loosely defined by Short *et al.* (2000) as a place for the transmission of economic, political, and cultural globalization. Such a broad definition may include all cities that might have any gateway opportunities to bring themselves into the global city network (essentially, the global business network). The gateway city concept in this sense is seen to be suitable for cities that would otherwise be unqualified to become a world city.

This discussion in the world city literature is a response to the discourse of world cities as a small group of elite places, rather than one based on the concept of intermediacy. In reality, however, the actor that drives to establish a city as a gateway of any kind is the same one that implements gateway-corridor development strategies – namely the state – although at different geographical scales.

As noticed by Jessop and Sum (2000), there is a growing trend of entrepreneurial cities at a global scale. In the Schumpeterian sense, they define and interpret the entrepreneurial city by identifying "five analytically distinct (but perhaps empirically overlapping) fields in which directly economic and/or economically relevant innovation can occur in relation to urban form and functions". These fields comprise:

1. The introduction of new types of urban place or space for producing, servicing, working, consuming, living, and so on;
2. New methods of space or place production to create location-specific advantages for producing goods/services or other urban activities;
3. Opening new markets – whether by place marketing specific cities in new areas and/or modifying the spatial division of consumption through enhancing the quality of life for residents, commuters, or visitors;
4. Finding new sources of supply to enhance competitive advantages; and
5. Refiguring or redefining the urban hierarchy and/or altering the place of a given city within it. Examples include the development of a world or global city position, regional gateways, hubs, cross-border regions, and "virtual regions" based on interregional cooperation among non-contiguous spaces. (Jessop and Sum 2000: 2290).

All the entrepreneurial efforts made in these fields by cities are considered to be "glurbanization", which differs from "glocalization" mainly in terms of identifying the strategic actors.

At the national level, there are various practices by which governments can proactively become involved in marketing, promoting, and coordinating the construction of gateways and corridors. Papers presented during roundtable conferences in Canada in 2007 (see www.gateway-corridor.com) are excellent sources on this topic.

It appears in the two papers that cover the overall situations of the gateway and corridor development in the USA (Goetz and Bandyopadhyay 2007) and Canada (Gillen *et al.* 2007) that these two federal governments have very comprehensive development plans or strategies, such as the National Corridor Planning and Development Program in the USA and the Asia-Pacific Gateway and Corridors Initiatives in Canada. These are considered as the *responses* to the new pattern of trade in North America (due largely to the North American Free Trade Agreement) and in Asia (largely caused by the world production bases in China and India).

Table 16.1 Glurbanization versus Glocalization

	Glurbanization	Glocalization
Strategic Actors	Cities (perhaps as national champions)	Firms (perhaps in strategic alliances)
Strategies	Place- and space-based strategies	Firm- or sector-based strategies
New scales of activities and temporalities	Create local differences to capture flows and embed mobile capital	Develop new forms of scalar and/or spatial division of labour
Chronotopic governance	Re-articulate time and/or systemic competitive advantages	Re-articulate global and local for dynamic advantages and space for structural advantages

Source: Adapted from Figure 1 (Jessop and Sum 2000).

On their geographical impacts, Rodrigue (2007) summarizes the new forms of gateways and corridors that reflect the imbalanced trade between Asia and North America. Meanwhile, Hall (2007b) points out the economic, social, and environmental impacts of super-sized logistics zones on the host gateway cities.

On the development opportunities of gateways and corridors, Button (2007: 15) points out that "the reduced role of government in terms of both active investment and regulation of operations means that the degree of uncertainty regarding the use, and the nature of the use, of any corridor is difficult to forecast". Similar arguments are offered by Vickerman (2007). When reviewing the European Union's (EU) series of Trans-European Networks (TENs), he states that "in this world there is no captive trade for individual networks", and that "the development of networks (and nodes) to compete for these global flows can have unpredictable regional impacts within countries" (Vickerman 2007: 20). Brooks (2007) identifies a significant infrastructure gap in the Atlantic gateway ports of Canada, the lack of funding and land for these ports being one of the causes. Heaver (2007: 10) also points out that:

> in general, strong leadership in the provision of gateway services, as distinct from investment in the port and promotion of a port gateway, has not been a mark of North American port authorities. Systems have worked well enough left to their own devices and levels of inter-port competition have not induced port authorities or governments to adopt more aggressive gateway strategies until recently.

The common ground underpinning these discussions is the role of the state – both as a regulator and a facilitator of the market. In fulfilling these responsibilities, the state must be very careful when choosing which areas it should promote. Typically, the choices are based largely on existing gateways, hubs, or corridors, making the state adaptive rather than proactive with respect to market needs or changes in global trade patterns. The real tasks, therefore, involve being able to upgrade these existing gateways, hubs and corridors and making them more competitive in meeting the new challenges of global supply chains and the like. In other words, states tend to help enhance rather than create geographical intermediacy (Fleming and Hayuth 1994).

However, such a condition may be unavailable not only in so-called maritime enclaves but also in coastal cities located in less favorable regions. These cities may become further disadvantaged if the hub-and-spoke port-call structure is established with more and frequent shipping services linking the major hub cities to the major foreland markets. Most importantly, market forces will not "correct" this spatial imbalance until the diseconomies of scale accumulate to a certain level in the existing hubs or major gateways or until a shift in trade relationships occurs. Moreover, the hub or gateway advantages for existing hub/gateway cities may prove their geographical centrality for trade-related activities and manufacturing area clusters.

What can a coastal city in a less favorable region of global trade do? Can it remain as a feeder to the hub/gateway in the region, or should it try to upgrade itself to a gateway city? From a market point of view, the latter is not an option; a city, no matter how entrepreneurial, cannot change the geography of trade in the region and its "inbetweenness". Who then could be the founders of new gateways or hubs in places that seem to lack intermediacy?

This chapter demonstrates that the state can play this role, but only in places where the regional government is strong and where there already exists a certain degree of centrality in the potential gateway city. The example presented is of gateway building by the entrepreneurial government in Guangxi, a less-developed province in China. After a brief introduction of the province's general situation, the main focus of this chapter is on the ways in which the provincial government has managed to integrate its coastal cities into an economic region and merge two ports into one gateway. In conclusion, the gateway concept in relation to China's behavior toward an entrepreneurial region is reviewed.

2. The Case of Guangxi, China

Guangxi Zhuang Autonomous Region (GZAR) is a provincial-level region in southern China (see Figure 16.1). It has 1,595 kilometers of coastline. It is land-bordered with Vietnam to the southwest; the rest of its land border is domestic, with Yunnan and Sichuan to the north and Guangdong Province to the east. Although Guangdong is considered the most economically dynamic region in China, the Pearl River Delta (PRD) and its core, Hong Kong, are some 900 kilometers away from the major port of Guangxi. Historically, Guangxi was considered a remote area in China. It was even called "Nan Man", which literally means the "boorish south". It did not have any port for industrial goods until the Vietnam War in the early 1960s when Port Fangchenggang was constructed to facilitate military and logistics supplies from China to North Vietnam. Since its establishment, Fangchenggang has become the only major port in the region. With a total designed berth handling capacity of 68.5 million tons, the port reached 33.8 million tons of throughputs, mainly bulk sulfur, coal, iron ore, grain, woodchips, and containers, in 2007. Among all major coastal ports in China, this may be the only port that is largely underutilized. The port plays a role of secondary gateway, after Zhanjiang in Guangdong, and only with respect to bulk cargo (coal and iron ore) for Sichuan, the largest Chinese inland province. For container shipping, it handles as few as 200,000 TEU a year, much less than most other coastal ports in China. This number is low because Guangxi is neither a major producer of, nor a market for, containerized products. From a port-city perspective, Fangchenggang City is also one of the smallest economies among all Chinese coastal cities with a major port.

Similar to its main port, the Autonomous Region has also ranked among the least developed coastal provinces. In 2007 its total gross provincial product (that is, the provincial GDP) was 595.6 billion Yuan, which ranked it ninth among ten coastal provinces of the mainland and was only marginally larger than the total gross provincial product of Tianjin Autonomous City. In terms of foreign trade,

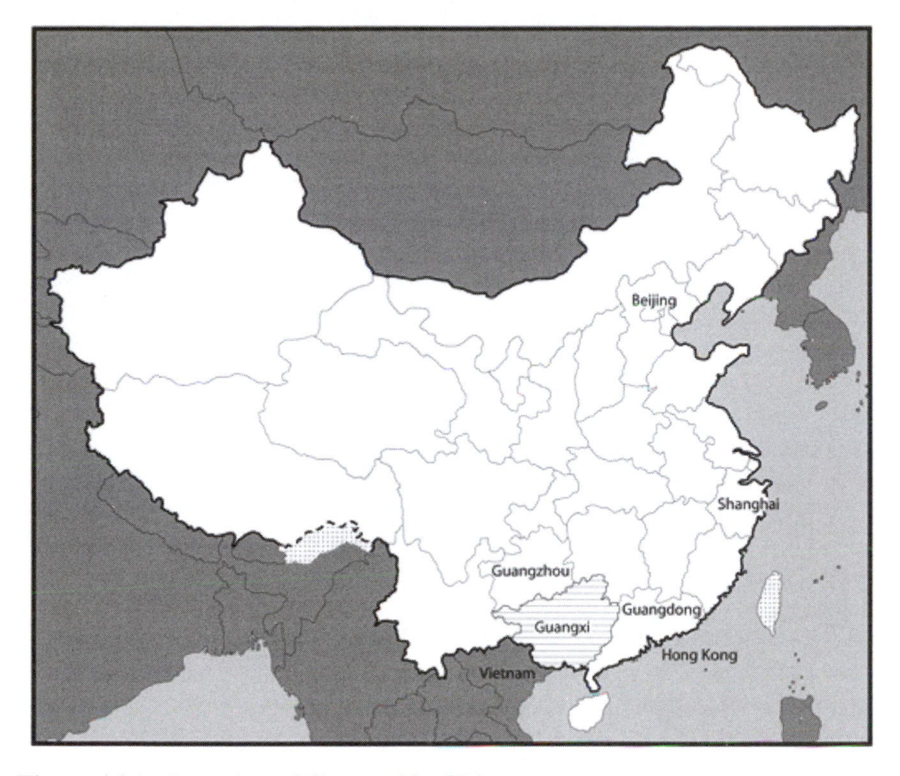

Figure 16.1 Location of Guangxi in China

the total export of Guangxi was only about 1 per cent of Shanghai or Guangdong in 2007, indicating a very low level of global linkage.

Table 16.2 provides a comparison of ten coastal provinces in China in terms of export values, total port throughput tonnages, and GDP in 1997 and 2007. These three indicators are used together to represent overall economic activity in relation to domestic and foreign trade through coastal or maritime flow of cargo. As indicated in Table 16.2, some provinces or principal level autonomous cities, such as Guangdong and Shanghai, have led for the past three decades. Some other provinces, like Shandong, have progressed well in both export and port throughput, indicating that they gained increasingly from international trade. Guangxi does not belong to either of these two situations. It has ranked at the bottom of all three indicators across the three time periods. Although its port throughput growth was above that of most of the other provinces in this ten year period, its exports in 2007 were only 21 per cent more than 1997, much lower than all other provinces.

These figures illustrate that in terms of economic scale, neither the GZAR nor its major port city of Fangchenggang incurred any advantage while competing with their counterparts in Guangdong and the other coastal provinces of China. With a

Table 16.2 Comparison of Chinese Coastal Provinces

Coastal provinces (from northeast to southwest)	1997 Export		1997 Throughput of coastal ports		1997 GDP		2007 Export			2007 Throughput of coastal ports			2007 GDP		
	Total (mil USD)	Rank	Total (000 tons)	Rank	Total (bil) Yuan	Rank	Total (mil USD)	As % of 1997	Rank	Total (000 tons)	As % of 1997	Rank	Total (bil) Yuan	As % of 1997	Rank
Liaoning	8.90	7	88,690	5	349	6	35.3	397	7	414,920	468	5	1,102	316	7
Hebei	7.38	8	78,620	6	395	5	17.0	230	9	399,840	509	6	1,371	347	5
TianjiTianjin	5.02	9	67,890	7	124	10	38.3	760	6	309,460	456	7	505	407	10
Shandong	13.1	4	113,770	4	665	3	75.2	575	5	575,470	506	2	2,597	391	2
Jiangsu	14.1	3	16,520	9	668	2	203.7	1,446	3	91,350	553	9	2,574	385	3
Shanghai	14.7	2	163,970	1	336	7	212.4	1,443	2	492,270	300	4	1,219	363	6
Zhejiang	11.1	6	121,870	3	464	4	76.8	691	4	574,390	471	3	1,878	405	4
Fujian	11.6	5	44,850	8	300	8	34.8	301	8	236,030	526	8	925	308	8
Guangdong	72.7	1	123,840	2	732	1	238.2	328	1	802,820	648	1	3,108	425	1
Guangxi	2.38	10	11,780	10	202	9	2.87	121	10	71,920	611	10	596	295	9

Source: Statistics Yearbooks of each province published by the Statistics Bureau of the respective provinces in various years.

limited accumulation of wealth and a less preferred location, Fangchenggang and the other two coastal cities in Guangxi (Qingzhou and Beihai) do not seem to be attractive for foreign direct investment (FDI), although all three city governments have been entrepreneurial in planning their port development aggressively. The fact remains that they could not attract new shipping lines in the past, and only recently has the investment atmosphere changed to their advantage.

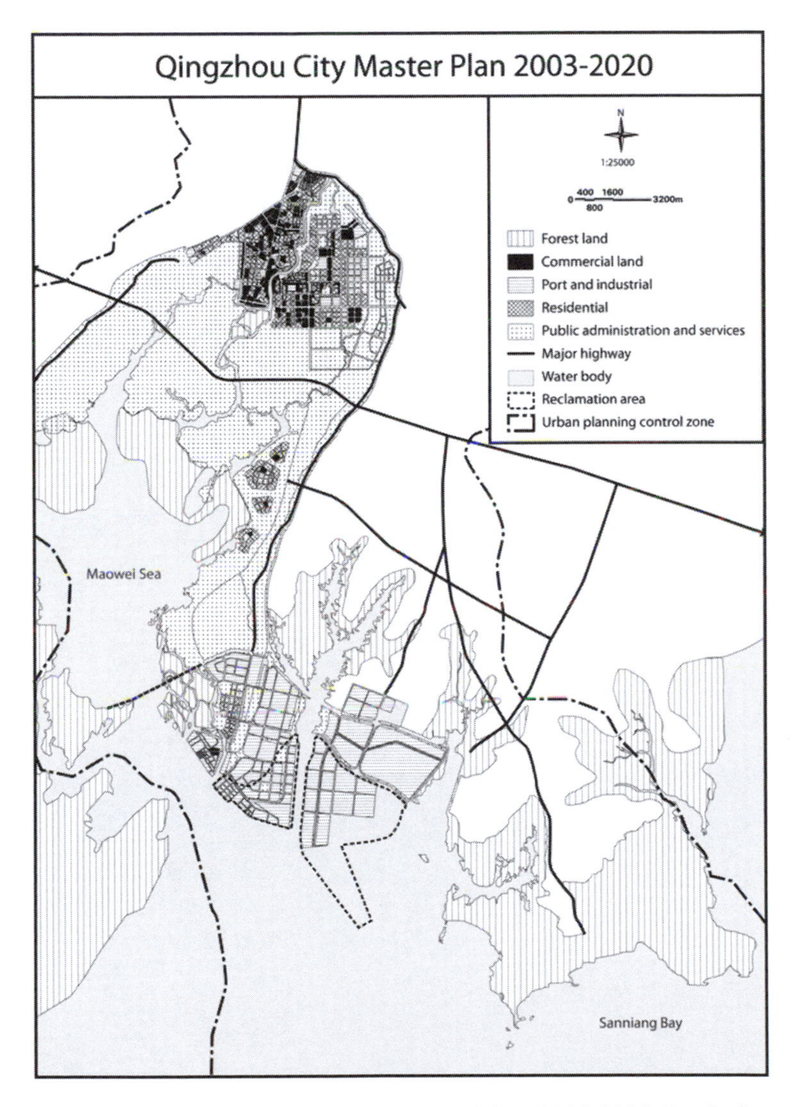

Figure 16.2 Qingzhou City Master Plan Revision (2003-2020, Draft, Scenario One)

Source: Chinese Academy of Urban Planning and Design.

The port city competition within Guangxi increased a few years ago with the development of Qingzou, the coastal city closest to the provincial capital. From 2006 to 2008, nine deepwater berths for large ships (from 70,000 to 150,000 tons) were built, and six deepwater channels with various sizes (for 50,000-ton to 200,000-ton class vessels) were dredged. A total of 13 billion Yuan was invested by both private sector and local government in a disaggregated approach for the three port cities. As a result, a total of 94 million tons of designed berth handling capacity was reached by the end of 2008, about 150 per cent of its total throughput for the same year. Figures 16.2 and 16.3 are the development plans of Qinzhou and Fangchenggang, respectively, indicating their ambition and determination.

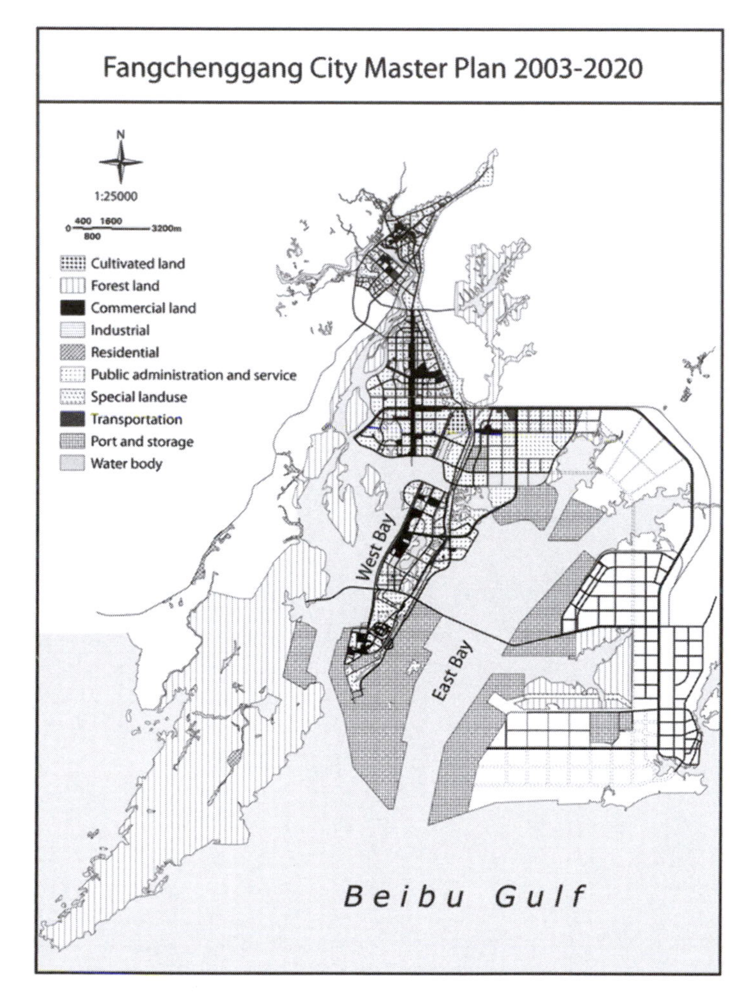

Figure 16.3 Fangchenggang City Master Plan (2003-2020, Draft)
Source: Chinese Academy of Urban Planning and Design.

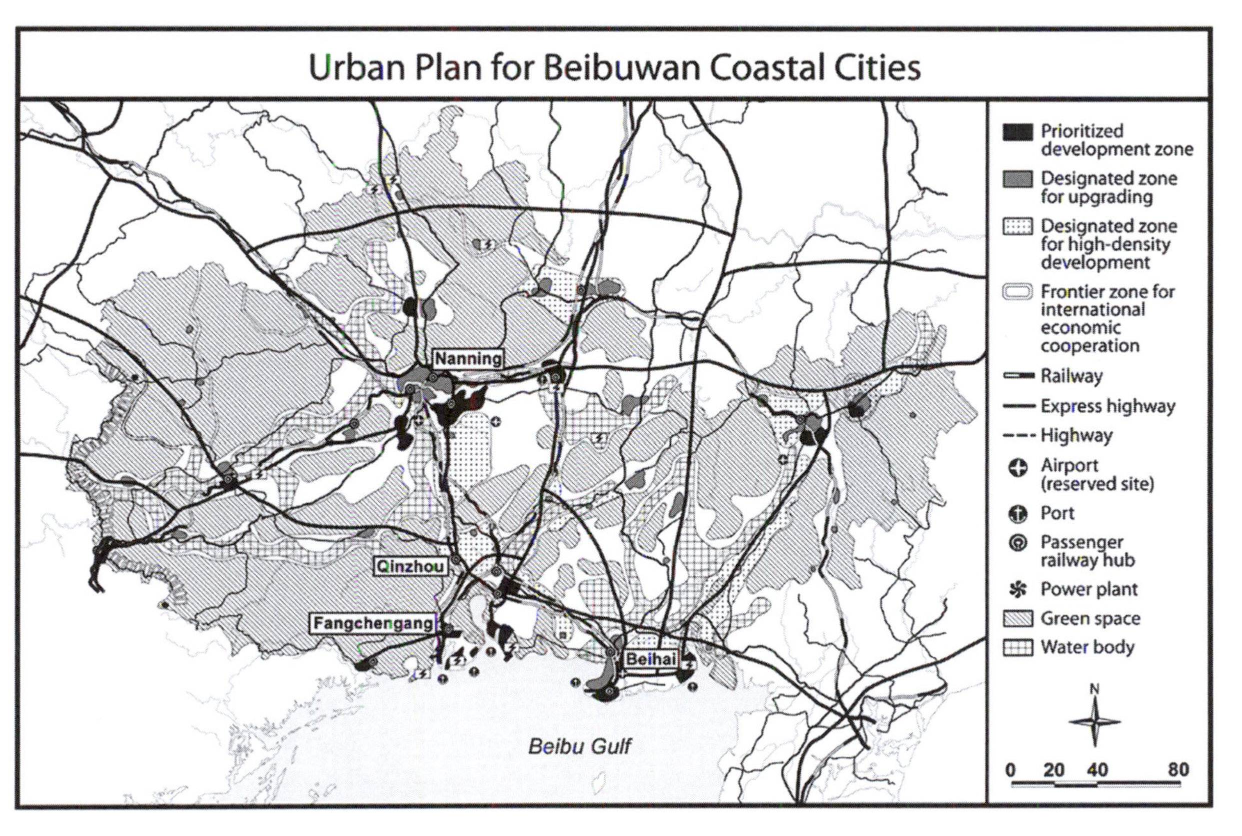

Figure 16.4 Urban Agglomeration Plan of Beibuwan Coastal Economic Region, Guangxi, China
Source: Chinese Academy of Urban Planning and Design.

The big transformation began in 2006 when the GZAR government decided to formulate a Nan(ning) – Bei(hai) – Qin(zhou) –Fang (chenggang) Coastal Economic Region (NBQFCER), which received approval by the central government at the beginning of 2008. It was eventually renamed the Beibuwan Coastal Economic Region (BCER) (see Figure 16.4). The BCER includes all three coastal cities, namely Fangchenggang, Qinzhou, and Beihai, and the provincial capital city of Nanning. The chairman of the BCER Steering Committee was the first deputy governor of the GZAR, implying the structure's administrative power over all the cities.

The BCER has 42,500 square kilometers of land and 12.5 million inhabitants. The overall objective of the BCER is to make a near-port-industrial cluster, an articulation space for the ASEAN-China Free Trade Agreement (ACFTA) cooperation.

The BCER is responsible for new state policies in the following six respects:

1. Administrative restructuring. This includes integrating three ports into one, the Beibuwan (Beibu Gulf) Port, effective March 2009. The Guangxi Beibu Gulf International Port Group (GBGIPG) that operates the new Beibuwan Port has further merged the railway company that provides services linking all the three ports to the national railway system.
2. Land-use incentives to attract investors, domestic or international.
3. Clustering of key projects. Table 16.3 lists some of the major investors and their projects under development since 2006. Although some of these are transnational corporations, all have to be guided in their business setup in this region; some may gain special financial incentives to invest in a specific place.
4. Policy incentives for infrastructure and transport and logistics services. For example, a 30 km long deepwater channel with a depth of 16 m in Qingzhou has been dredged and was completed in 2008. From the total dredging cost of one billion Yuan (about USD 135 million), the provincial government paid one-third and the central government paid one-fifth. A new 50 km long canal is being planned and is ready to be built by 2020 to connect the capital city Naning to the Beibu Wan Port, along with new highways and railway links. Another area of incentive is the subsidization of transport costs. On the land side, all container trucks going to or from the ports in the BCER can avail of a 50 per cent deduction in all port traffic charges, while all vehicles registered in the BCER can enjoy a 20-35 per cent deduction of toll charges (GXDRC 2008). On the sea side, the provincial government provides subsidy through BCER to operate four new feeder services to connect Beibuwan Port to Hong Kong.
5. Policies for banking and financing. The establishment of local banks and the issuing of bonds are encouraged. Beihai Port, being one of the four companies merged into the GBGIPG, is a publicly listed company in the Shenzhen Stock Market. The GBGIPG has become the largest shareholder

Table 16.3 Selected Projects Under Construction or Being Planned in the Beibuwan Coastal Economic Region (BCER)

Name of the Company	Industry Type	Capacity	Investment Scale	Location within the BCER	Completion Time (planned)
Petrol China	Oil refinery plant	10 mil tons/year or 200,820 barrels/day	15 Billion Yuan (about USD 2.2 billion)	Qinzhou Bonded Port Area	Second half of 2009
APP (Asia Pulp and Paper Co., Ltd.) (Indonesia)	Pulp and Paper	1.8 million tons of pulp and 3.1 million tons of paper/year	41 Billion Yuan (about USD 6 billion)	Qingzhou	Phase One (2009) Phase Two (2012)
Wuhan Iron and Steel Co. and Liuzhou Iron and Steel Co.	Iron and Steel	10 million tons/year	68.6 Billion Yuan (about USD 10 Billion)	Fangchenggang	2012
Stora Enso (Finland)	Integrated Forestry, Pulp, and Paper	900,000 tons of pulp and 900,000 paper/year (Phase One)	22.7 Billion Yuan (about USD 3 Billion)	Beihai	The end of 2010
Guangdong Nuclear Power Group, China Power Investment Co. and Guangxi Nuclear Power Group	Nuclear Power Station	2 million KW (Phase One); 6 million KW (final)	25 Billion Yuan (about USD 2.8 Billion) for Phase One	Fangchenggang	Undecided, the project has not yet been started as of May 2009
The BCER	**All Major Infrastructure Projects**		**40 billion Yuan (about USD 5.8 billion)**	**Beibu Gulf Economic Region**	**Up to March 2009**

Source: Compiled by the author, based on information gathered from various government websites.

after assuming all the assets of the other three companies. This not only reduces the risk to the GBGIPG, but also provides new channels to finance its further development.

6. Incentives for international and interregional cooperation. The BCER is recognized by the Chinese central government as "the hub that links China with ASEAN countries", and is therefore a place specially designed for ACFTA countries. For this reason, the central government approved in May 2008 the opening of the sixth Bonded Port Area (BPA) in the country, with ten square kilometers of land including the port of Qingzhou. The term "bonded port area" refers to areas under special custom supervision that have the functions of ports, logistics, and processing, among others. They are created upon approval of the State Council, and have been established within or adjacent to port areas. Customs houses supervise the means of transport of goods and articles entering and exiting the BPAs, as well as the enterprises and locations within them. The State Council approved the establishment of the first BPA in June 2005 (the Shanghai Yangshan Bonded Zone). Subsequently, approval was given to set up Tianjin Dongjiang, Dalian Dayaowan, Hainan Yangpu, Ningbo Meishan, Guangxi Qinzhou, and Xiamen Haicang – a total of six BPAs – to achieve the combination of port logistics capabilities and the special policies of free trade zones.

Intensive investment in port and industries has been reflected in local statistics. In 2008, and even in the first part of 2009, when the world economy was in deep recession, BCER had a record-high GDP and port throughput. Indeed, these projects provide sound support for BCER and port development under a certain level of centrality, such that the coastal region does not need to rely heavily on its gateway functions.

3. Discussion on the Entrepreneurial Region

Although it is still too early to assess the performance of BCER as a growth pole and the Beibuwan Port as a new gateway between China and ASEAN countries, discussions still abound in relation to this new gateway region.

The BCER is a case of an entrepreneurial region where the state is an actor in making and promoting a "place" as large as a group of four cities. More importantly, we may interpret the whole action as a strategic move of a weak province in a global and/or national competition with other actors who still function at the city level. This may not be possible in many countries; in China, however, such state-driven development has prevailed for decades.

In the case of BCER, port competition and spatial evolution of port regionalization that are driven by market forces have been intercepted by a proactive state. The merger of three ports in BCER is similar to a horizontal integration of a port operator. It can be defined as a spatial integration by an entrepreneurial region.

As the enactment of this spatial integration has occurred before extensive port and city competition within the region due to the behavior of an entrepreneurial city, the merger can be interpreted as an internalization of production to reduce the transaction costs when producing port services by the province as a whole.

The inclusion of the capital city of Nanning as a leading part of BCER, and the plan to connect it by a new canal, reveals an intention to coastalize the core development of Guangxi. Although it is too early to say that such a move is a success, the process of this gateway-building is similar to the models developed by Rimmer (1977) and Taaffe *et al.* (1963) based on the situations of colonial South East Asia and West Africa respectively, where capital cities of the colonies moved outward to the coastal port cities. However, the driving force behind the BCER is very different from these examples; rather, it is a spatial strategy of an entrepreneurial region to gain gateway status in a globalizing world.

The Guangxi case also differs from gateway- and corridor-building in the USA in terms of timing. Blank *et al.* (2007) point out that since NAFTA operations in 1991, many working groups for NAFTA improved many aspects of cross-border trade and related issues. However, other than border infrastructure, little attention was given to larger issues of North American transport infrastructure and on early post-NAFTA thinking about the creation of NAFTA Super Highways. In the case of Guangxi, we note that the ACFTA was signed in January 2007 and will be implemented in 2010. The building of BCER and all related gateway infrastructure is in fact a few years ahead of the ACFTA implementation. In this regard, we may argue that the US model of gateway development is adaptive to trade patterns while the Chinese approach is proactive. Such an approach, which is carried out at a regional scale, should be seen as a representative behavior of an entrepreneurial region.

In China, Guangxi is but only one of many cases of entrepreneurial regional development. For example, in Hebei, another poor coastal province without its capital focused on coastal resources, a new city-region of more than one million inhabitants has built a new seaport on Caofeidan, an island 18 km away from the seashore.

4. Conclusions

We started with a brief review of the gateway concept and the "glurbanization" perspective before raising the following research question: is it possible for a coastal city with a less favorable location in the global arena to establish itself through proactive gateway building? By examining the case of Guangxi, we find that such a task of glurbanization may be carried out by an entrepreneurial region instead of individual entrepreneurial cities.

From a transport geography point of view, the case of Guangxi illustrates that the provincial government may be able to mobilize resources at a regional level with a geographical focus toward becoming a gateway zone. With the strong support of the Guangxi provincial government and the Chinese national government, it seems possible for the BCER to build gateway infrastructure and offer services while simultaneously building economic centrality.

Chinese gateway-making models, however, may be difficult for other developing countries, or countries with a democratic political system, to emulate for the following three reasons. First, this model needs proactive rescaling in order to reduce the "territorial" transaction cost and to achieve geographical efficiency in gateway- and corridor-building. Such proactive rescaling is not easy to achieve in a democratic society or a country lacking long-term political stability. Second, unless a region is located in a development-driven economy led strongly by the state, it would be very difficult to mobilize resources from different geographical levels, from national to local, in order to make a gateway-corridor region ahead of market demand. Third, in terms of social concerns, the Chinese model of building gateways in coastal greenfield locations may be justified as being able to absorb a large amount of the surplus rural agricultural labor that has been rapidly urbanizing since the 1980s. The large-scale coastalization of productive activities in many other countries may no longer hold true, such as in Japan and Korea where the peak period of urbanization and industrialization is over.

When putting China's entrepreneurial region into a global context, one finds that the case of Guangxi BCER may be a supply-led gateway development or a proactive rescaling approach to fulfilling the resource and power requirements for development. It may also fit the needs of those on the producer side of global trade. Indeed, China has shown very different spatial patterns of development around port areas (Rodrigue and Notteboom 2007). Almost all major hub ports in China have had a spatial leap forward to a new port area built in a distant place 50 to 100 km away from their cities.

In analyzing the building of BCER as a spatial reconstruction of a city-region, one would note that this has been discussed among scholars, such as Hall (2007a) and O'Connor (2008) when examining logistics zones of global scale. The BCER in Guangxi, and similar coastal economic regions in China and other developing countries that have become more export-dependent over recent years such as India and Brazil, rely on and may expect positive results from the migrant workforce from other regions within their own countries. Therefore, the process of such gateway- or logistics zone-building is in fact the making from scratch of new global logistics-based city-regions. Problems and impacts of this process may be very different from the situation in mature cities, such as in Vancouver, Canada (Hall 2007b; Heaver 2007). In other words, the case of the Guangxi BCER characterizes the strategies that an entrepreneurial region may carry out during glurbanization in the developing world. The comprehensive discourse and interpretation of state-market interplay in the regional/global gateway place-making process is an important area for further research in order to gain a better understanding of the urban-region geographies of global trade, shipping, and logistics.

Bibliography

Abell, D.F. and Hammond, J.S. 1979. *Strategic Market Planning: Problems and Analytical Approaches*. Englewood Cliffs, NJ: Prentice-Hall.

African Development Bank. 2009. *Africa and the Global Economic Crisis: Strategies for Preserving the Foundations of Long-term Growth*. Working Paper 98, prepared for 'The 2009 Annual Meetings of the African Development Bank': May 13-14 2009, Dakar, Senegal.

Alix, Y. and Germain, O. 2008. *Toward a Generic Business Strategy in the Liner's Industry: Discussion on World Port Development*. Proceedings of the 24th International Port Conference, Port Training Institute, Alexandria, Egypt, 18-20 February 2008.

Alix, Y., Slack, B. and Comtois, C. 1999. Alliance or acquisition? Strategies for growth in the container shipping industry: The case of CP Ships. *Journal of Transport Geography*, 7(3), 203-8.

Allison, P.D. 1978. Measures of inequality. *American Sociological Review*, 43, 865-80.

Alphaliner. 2010. *Weekly Newsletter*, 7.

Anderson, E.W. and Fornell, C. 2000. Foundations of the American customer satisfaction index. *Total Quality Management*, 11(7), 869-82.

Arvis, J-F., Mustra, M.A., Panzer, J., Ojala, L. and Naula, T. 2007. *Connecting to Compete: Trade Logistics in the Global Economy – The Logistics Performance Index and its Indicators*. New York, NY: The World Bank.

Ashar, A. 1999. The fourth revolution. *Containerisation International*, December, 57-61.

Associated British Ports. 2009a. *ABP Connect*. Available at: http://www.abports.co.uk/custinfo/other_op/abp_connect.htm [accessed: September 2009].

Associated British Ports. 2009b. *Exxtor Shipping Services*. Available at: http://www.abports.co.uk/custinfo/other_op/exxtor.htm [accessed: September 2009].

Associated British Ports. 2009c. *ABP.* Available at: www.abports.co.uk [accessed: September 2009].

Australia Bureau of Infrastructure, Transport, Regional Economics (Australia-BITRE). 2008a. *Waterline*, 43. Canberra: Department of Infrastructure, Transport and Regional Economics.

Australia Bureau of Infrastructure, Transport, Regional Economics (Australia-BITRE). 2008b. *Waterline*, 44. Canberra: Department of Infrastructure, Transport and Regional Economics.

AXS-Alphaliner. 2009. *AXSMARINE: Connecting the Shipping Community*. Available at: www.axsmarine.com [accessed: 27 April 2009].

Baird, A.J. 2002. The economics of container transhipment in northern Europe. *International Journal of Maritime Economics*, 4(3), 249-80.

Ballou, R.H. 1992. *Business Logistics Management*. 3rd Edition. London: Prentice Hall International.

Ballou, R.H. 2007. The evolution and future of logistics and supply chain management. *European Business Review*, 19(4), 332-48.

Baltazar, R. and Brooks, M.R. 2007. Port governance, devolution and the matching framework: A configuration theory approach, in *Devolution, Port Governance and Port Performance*, edited by M.R. Brooks and K. Cullinane. London: Elsevier, 379-403.

Baltic Exchange. 2009. Available at: http://www.balticexchange.com [accessed: May 2009].

Barber, E. 2008. How to measure the "value" in value chains. *International Journal of Physical Distribution and Logistics Management*, 38(9), 685-98.

Barros, C.P. 2003. Incentive regulation and efficiency in Portuguese seaports. *Maritime Economics and Logistics*, 5(1), 55-69.

Barros, C.P. 2006. A benchmark analysis of Italian seaports using data envelopment analysis. *Maritime Economics and Logistics*, 8, 347-65.

Barros, C.P. and Athanasiou, M. 2004. Efficiency in European seaports with DEA: Evidence from Greece and Portugal. *Journal of Maritime Economics and Logistics*, 6(2), 123-40.

Basole, R.C. and Rouse, W.B. 2008. Complexity of service value networks: Conceptualization and empirical investigation. *IBM Systems Journal*, 47(1), 53-70.

Beamon, B.M. 1999. Measuring supply chain performance. *International Journal of Operations and Production Management*, 19(3), 275-92.

Beaverstock, J.V., Doel, M.A., Hubbard, P.J. and Taylor, P.J. 2002. Attending to the world: Competition, cooperation and connectivity in the World City Network. *Global Networks*, 2(2), 111-32.

Beresford, A.K.C., Gardner, B.M., Pettit, S.J., Naniopoulos, A. and Wooldridge, C.F. 2004. The UNCTAD and WORKPORT models of port development: Evolution or revolution. *Maritime Policy and Management*, 31(2), 93-107.

Beresford, A.K.C., Gardner B.M., Pettit, S.J., Wooldridge C.F. and Naniopoulos, A. 2000. *Synthesis of Results Concerning New Organisational Structures and Suggestions for the Transitional Process*. WORKPORT (WA-97-SC-2213). Deliverable 7 for European Commission DG TREN, University of Wales, Cardiff, Department of Maritime Studies and International Transport. December 1999, 58.

Bichou, K. 2007. Review of port performance approaches and a supply chain framework to port performance Benchmarking, in *Devolution, Port Governance and Port Performance*, edited by M.R. Brooks and K. Cullinane. Amsterdam: Elsevier, 567-98.

Bichou, K. and Gray, R. 2004. A logistics and supply chain management approach to port performance measurement. *Maritime Policy and Management*, 31(1), 47-67.

Bird, J. 1963. *The Major Seaports of the United Kingdom*. London: Hutchison.

Blank, S. 2006. *North American Trade Corridors: An Initial Exploration in Notes and Analyses on the USA*. Montréal: Université de Montréal.

Blank, S., Golob, S. and Stanley, G. 2007. *A North American Transportation Infrastructure Strategy*. Paper to the Winnipeg Roundtable, Asia-Pacific Gateway and Corridor Research Consortium, Winnipeg, Canada, 21-22 February 2007. Available at: www.gateway-corridor.com [accessed: 22 May 2009].

Blenkey, N. 2009. What's blue, green and wet: Marine highway. *Marine Log*, 23-25 February.

Blumenhagen, D. 1981. Containerisation and hinterland traffic. *Maritime Policy and Management*, 8, 197-206.

Bonacich, E. and Wilson, J.B. 2007. *Getting the Goods*. Ithaca, NY: Cornell University Press.

Bots, P. and Lootsma, F.A. 2000. Decision support in the public sector. *Journal of Multi-Criteria Decision Analysis*, 9(1), 1-3.

Bour, N. 2008. The Seine-Nord Europe canal: Central segment of the Seine-Scheldt water-way link. *PIANC Magazine*, 132, 5-11.

Bourne, M., Mills, J., Wilcox, M., Neely, A. and Platts, K. 2000. Designing, implementing and updating performance measurement systems. *International Journal of Operations and Production Management*, 20(7), 754-71.

Brooks, M.R. 2000. *Sea Change in Liner Shipping: Regulation and Managerial Decision-making in a Global Industry*. Oxford: Elsevier.

Brooks, M.R. 2007. *Gateways and Canada's Ports Policy: Issues and Impediments*. Paper to the Vancouver International Conference, Asia-Pacific Gateway and Corridor Research Consortium: Vancouver, Canada, 2-3 May 2007. Available at: www.gateway-corridor.com [accessed: 25 May 2009].

Brooks, M.R. and Pallis, A.A. 2008. Assessing port governance models: Process and performance components. *Maritime Policy and Management*, 35(4), 411-32.

Brooks, M.R, McCalla, R., Pallis, A. and Van der Lugt, L. 2009. *Coordination and Cooperation in Strategic Port Management: The Case of Atlantic Canada's Ports*. Paper to the International Association of Maritime Economists Conference, Copenhagen. *Containerisation International Yearbook* (various). London: Informa.

Bruyas, F. 2000. Port Saïd (Egypte), lieu d'articulation du local ou mondial. Zone et ville franche: Questions d'échelles. *Annales de Géographie*, 6(12), 152-71.

Burghardt, A.F. 1971. A hypothesis about gateway cities. *Annals of the Association of American Geographers*, 61(2), 269-85.

Buttler, F. 1975. *Growth Pole Theory and Economic Development*. Westmeath: Saxon House.

Button, K. 2007. *Distance and Competitiveness – Emerging Continental Network Barriers and Strategic Partners*. Paper to the Winnipeg Roundtable, Asia-Pacific Gateway and Corridor Research Consortium,Winnipeg, Canada, 21-22 February 2007. Available at: www.gateway-corridor.com [accessed: 22 May 2009].

Byrne, B.M. 2001. *Structural Equation Modeling with AMOS Basic Concepts, Applications, and Programming*. Mahwah, NJ: Lawrence Erlbaum Associates.

Campbell, S. 1993. Increasing trade, declining port districts: Port containerization and the regional diffusion of economic benefits, in *Trading Industries, Trading Region: International Trade, American Industry and Regional Economic Development*, edited by H. Noponen, J. Graham and A.R. Markusen. New York, NY: Guilford Press, 212-55.

Canada. 2006. *Prime Minister Harper Launches Asia-Pacific Gateway and Corridor Initiative*. Available at: http://www.pm.gc.ca/eng/media.asp?id=1352 [accessed: 30 October 2009].

Canadian Chamber of Commerce. 2008. *Moving the Canadian Economy – Four Pillars For a National Transportation Strategy*. Available at: http://www.chamber.ca/images/uploads/Reports/transportation-strategy.pdf [accessed: March 2010].

Canadian Press. 2008. N.S. container terminal gets environmental OK. 23 October 2008.

Canadian Shipowners Association. 2008. *Annual Report 2008*.

Carbone, V. and De Martino, M. 2003. The changing role of ports in supply-chain management: An empirical analysis. *Maritime Policy and Management*, 30(4), 305-20.

Carter, C.R. and Jennings, M.M. 2002. Social responsibility and supply chain relationships. *Transportation Research Part E*, 38, 37-52.

Charlier, J. 1983. Ports et régions françaises: Une analyse macrogéographique. *Acta Geo-graphica Lovaniensia*, 24, 1-198.

Charlier, J. 1985. L'évolution récente du système portuaire canadien face à la conteneurisation et l'émergence de Montréal en tant que porte du Canada central et du Middle-West américain pour le trafic conteneurisé transatlantique. *Le Journal de la Marine Marchande*, 3420, 1595-609.

Charlier, J. 1991. L'arrière-pays national du port du Havre. *L'Espace Géographique*, 19-20, 325-34.

Charlier, J. 1992. Ports and hinterland connections, in *Ports as Nodal Points in a Global Transport System*, edited by A. Dolman and J. Van Ettinger. Oxford: Pergamon, 105-21.

Charlier, J. and Ridolfi, G. 1994. Intermodal transportation in Europe: Modes, corridors and nodes. *Maritime Policy and Management*, 21(3), 237-50.

Cho, J. and Kang, J. 2001. Benefits of challenges of global sourcing: Perceptions of US apparel retail firms. *International Marketing Review*, 18, 542-60.

Christopher, M. 1998. Relationships and alliances – Embracing the era of network competition, in *Strategic Supply Chain Management Alignment –*

Best Practice in Supply Chain Management, edited by J. Gattorna. Aldershot: Gower, 273-84.

Christopher, M. 2005. *Logistics and Supply Chain Management*. 3rd Edition. London: Prentice Hall.

Christopher, M., Peck, H. and Towill, D. 2006. A taxonomy for selecting global supply chain strategies. *International Journal of Logistics Management*, 17(2), 277-87.

Chung, K-C. 1993. *Port Performance Indicators*. New York, NY: The World Bank.

Coe, N.M., Dicken, P. and Hess, M. 2008. Global production networks: Realizing the potential. *Journal of Economic Geography*, 8(3), 271-95.

Coe, N.M., Hess, M., Yeung, H.W-C., Dicken, P. and J. Henderson. 2004. Globalizing regional development: A global production networks perspective. *Transactions of the Institute of British Geographers*, 29, 468-84.

Council of Ministers Responsible for Transportation and Highway Safety. 2002. *A Partnering for the Future: A Transportation Vision for Canada*.

Council of Ministers Responsible for Transportation and Highway Safety. 2008. *Harmonization of Transportation Policies and Regulations: Context, Progress and Initiatives in the Motor Carrier Sector*. Available at: http://www.comt.ca/english/coff-report.pdf [accessed: March 2010].

Croxton, K.L., Garcia-Dastugue, S.J., Lambert, D.M. and Rogers, D.S. 2001. The supply chain management processes. *The International Journal of Logistics Management*, 12(2), 13-36.

Cuadrado, M., Frasquet, M. and Cervera, A. 2004. Benchmarking the port services: A customer oriented proposal. *Benchmarking: An International Journal*, 11(3), 320-30.

Cudaly, B. 2006. *Box Boats: How Containerships Changed the World*. New York, NY: Fordham University Press.

Cullinane, K. and Khanna, M. 2000. Economies of scale in large containerships: Optimal size and geographical implications. *Journal of Transport Geography*, 8, 181-95.

Cullinane, K., Khanna, M. and Song, D-W. 1999. *How Big is Beautiful? Economies of Scale and the Optimal Size of Containerships*. Proceedings of the IAME 1999 Conference, 'Liner Shipping: What's Next?', Halifax, 108-40.

Cullinane, K., Song, D-W. and Gray, R. 2002. A stochastic frontier model of the efficiency of major container terminals in Asia: Assessing the influence of administrative and ownership structures. *Transportation Research Part A*, 36, 743-62.

Cullinane, K., Wang, T.F., Song, D.W. and Ping, J. 2006. The technical efficiency of container ports: Comparing data envelopment analysis and stochastic frontier analysis. *Transportation Research Part A*, 40(4), 354-74.

De Langen, P.W. 2004. Governance in seaport clusters. *Journal of Maritime Economics and Logistics*, 6(2), 141-56.

De Langen, P.W. 2008a. *Port Performance Indicators*. Paper presented to the European Sea Ports Organisation, Bilbao, 24 September 2008.

De Langen, P.W. 2008b. *Ensuring Hinterland Access: The Role of Port Authorities*. Paris: Joint Transport Research Centre of the OECD and the International Transport Forum, Discussion Paper No. 2008-11.

De Langen, P.W., Nijdam, M. and Van der Horst, M. 2007. New indicators to measure port performance. *Journal of Maritime Research*, 4(1), 23-36.

De Martino, M. and Morvillo, A. 2008. Activities, resources and inter-organizational relationships: Key factors in port competitiveness. *Maritime Policy and Management*, 35(6), 571-89.

De Monie, G. 1987. *Measuring and Evaluating Port Performance and Productivity*. Geneva: UNCTAD.

De, P. and Ghosh, B. 2003. Causality between performance and traffic: An investigation with Indian ports. *Maritime Policy and Management*, 30(1), 5-27.

Debrie, J. 2001. *De la continentalité à l'Etat enclavé. Circulation et ouverture littorales des territoires intérieurs de l'ouest africain*. Thèse, Université du Havre, Centre interdisciplinaire de Recherches en Transports et Affaires internationales.

Debrie, J. 2004. Ports secs, intérieurs ou avancés: Réorganisation des arrière-pays portuaires ou concept publicitaire? *Transports*, 427, 300-306.

Debrie, J. and Guerrero, D. 2008. (Re)spatialiser la question portuaire: Pour une lecture géographique des arrière-pays portuaires européens. *L'Espace Géographique*, 37, 45-56.

Debrie J., Lavaud-Letilleul, V. and Soppé, M. 2008. La réforme des ports italiens: Éléments de réflexion pour la politique portuaire française. *Transports*, 447, 11-20.

Declaration. 2006. Declaration Between the Department of Transport of Canada, the Secretariat of Transport and Communications of the United Mexican States and the Department of Transportation of the United States of America. Vancouver. Available at: http://www.tc.gc.ca/policy/acf/shortseas/namc2006/declaration.htm [accessed: 12 May 2009].

Denktas, G. and Beresford, A.K.C. 2009. *Analysis of Multimodal Transport Alternatives between Turkey and the United Kingdom*. Proceedings of the 14th Logistics Research Network Conference, Cardiff, UK, 9-11 September 2009.

Derudder, B. and Taylor, P.J. 2005. The cliquishness of world cities. *Global Networks*, 5(1), 71-91.

Dicken, P. 1998. *Global Shift: Transforming the World Economy*. New York and London: Guilford Press.

Drewry Shipping Consultants. 2001. *Post-Panamax Containerships – The Next Generation*. London.

Drewry Shipping Consultants. 2008. *Annual Review of Global Container Terminal Operators*. London.

Ducruet, C. and Lee, S.W. 2006. Frontline soldiers of globalization: Port-city evolution and regional competition. *GeoJournal*, 67(2), 107-22.

Ducruet, C., Notteboom, T.E. and De Langen, P.W. 2009. Revisiting inter-port relationships under the new economic geography research framework, in *Ports in Proximity – Competition and Coordination among Adjacent Seaports*. Farnham: Ashgate, 11-27.

Durveaux, H. 2004. Port logistics: The Antwerp case, in *Proceedings of the First International Conference on Logistics Strategy for Ports*, edited by T. Notteboom and S. Lin. Dalian: Dalian Maritime University Press, 394-407.

Edmonson, R.G. 2009. Great lake expectations: Some investment and shipper needs may add up to a short-sea strategy for the St. Lawrence. *The Journal of Commerce Magazine*. March 30.

Eggert, A. and Ulaga, W. 2002. Customer perceived value: A substitute for satisfaction in business markets? *Journal of Business and Industrial Marketing*, 17(2/3), 107-18.

ESCAP. 2003. *Commercial Development of Regional Ports as Logistics Centres*. Economic and Social Commission for Asia and the Pacific. New York: United Nations.

Estache, A. and Trujillo, L. 2009. Global economic changes and the future of port authorities, in *Future Challenges for the Port Shipping Sector*, edited by H. Meersman, E. Van de Voorde and T. Vanelslander. London: Informa, 69-87.

European Commission. 2001. *European Transport Policy for 2010 – Time to Decide*. White Paper [COM(2001) 370]. Available at: http://ec.europa.eu/transport/strategies/doc/2001_white_paper/lb_texte_complet_en.pdf [accessed: 16 January 2010].

European Commission. 2007. *The EU's Freight Transport Agenda: Boosting the Efficiency, Integration and Sustainability of Freight Transport in Europe* [COM(2007) 606 final]. Available at: http://eur-lex.europa.eu/LexUriServ/LexUriServ.do?uri=COM:2007:0606:FIN:EN:PDF [accessed: 16 January 2010].

Evans, P. 2008. The Asia pacific gateway and the reconfiguration of North America. *Canadian Political Science Review*, 2(4), 93-8.

Everret, S. 2005. Policy making and planning for the port sector: Paradigms in conflict. *Maritime Policy and Management*, 32(4), 347-62.

Fabbe-Costes, N., Jahre, M. and Rouquet, A. 2006. Interacting standards: A basic element in logistics networks. *International Journal of Physical Distribution and Logistics Management*, 36(2), 93-112.

Farrell, S. 2009. *Factors Influencing Port Efficiency – A Case Study of Dar Es Salaam*. (Proceedings: CD-Rom). Paper to the International Association of Maritime Economists (IAME) Conference, Copenhagen, 24-26 June 2009.

Farris, M.T. 1997. Evolution of academic concerns with transportation and logistics. *Transportation Journal*, 37(1), 42-50.

Fawcett, S.E., Calantone, R.J. and Smith, S.R. 1996. An investigation of the impact of flexibility on global market. *International Journal of Physical Distribution and Logistics Management*, 30(6), 472-99.

Ferrari, C., Parola, F. and Morchio, E. 2006. Southern European ports and the spatial distribution of EDCs. *Journal of Maritime Economics and Logistics*, 8(1), 60-81.

Fisher, M. 1997. What is the right supply chain for your product? *Harvard Business Review*, March/April.

Fleming, J. and Hayuth, R. 1994. Spatial characteristics of transportation hubs: Centrality and intermediacy. *Journal of Transport Geography*, 2(1), 3-18.

Foggin, J. and Dicer, G. 1984. Disappearing hinterlands: The impact of the logistics concept on port competition. *Annual Proceedings of the Transportation Research Forum*, 15, 385-9.

Ford, D. 1980. The development of buyer-seller relationships in industrial markets. *European Journal of Marketing*, 14(5/6), 339-53.

Franc, P. 2007. Intérêts et rentabilité des dessertes terrestres massifiées pour les armements de lignes régulières conteneurisées. *Cahiers Scientifiques du Transport*, 52, 119-42.

Franc, P. and Van der Horst, M.R. 2008. *Analyzing Hinterland Service Integration by Shipping Lines and Terminal Operators in the Hamburg-Le Havre Range*. Proceedings of the Annual Conference of the Association of American Geographers, Boston, Massachusetts, USA, April 2008.

Frederick, B. 2008. Private sector make recommendations for St. Lawrence-Great Lakes gateway. *Canadian Sailings*, December 18.

Frémont, A. 2007a. Global maritime networks: The case of Maersk. *Journal of Transport Geography*, 15(6), 431-42.

Frémont, A. 2007b. *Le Monde en Boîtes*. Arcueil (Paris): Institut National de la Recherche sur les Transports et leur Sécurité (INRETS), *Synthèse*, 53.

Frémont, A. 2008. Les armements de ligne régulière et la logistique. *Cahiers Scientifiques du Transport*, 53, 123-43.

Frémont, A. 2009a. Shipping lines and logistics. *Transport Reviews,* 29(4), 537-54.

Frémont, A. 2009b. *Flux maritimes internationaux et desserte de l'Ile-de-France: un levier de l'intégration territoriale pour le Bassin parisien?* Presentation to the Seminar Le Bassin parisien: Un espace métropolitain: Paris, IAU Idf/UMR.Géographie-Cités. Available at : http://www.iaurif.org/fileadmin/user_upload/Enjeux/Bassin_parisien/seminaire2/PPT_Fremont.pdf [accessed: 28 April 2009].

Frémont, A. and Soppé, M. 2007. Northern European range: Shipping line concentration and port hierarchy, in *Ports, Cities and Global Supply Chains*, edited by J. Wang, D. Olivier, T. Notteboom and B. Slack. Aldershot: Ashgate, 105-20.

Frémont, A., Franc, P. and Slack, B. 2008. Desserte fluviale des ports maritimes et transport des conteneurs: Quels enjeux pour les ports français du Havre et de Marseille dans le contexte européen? *Cybergeo*, 437, 1-21.

Frémont, A. and Lavaud-Letilleul, V. 2009. Rethinking proximity: New opportunities for port development – The case of Dunkirk, in *Ports in*

Proximity: Competition and Coordination Among Adjacent Seaports, edited by C. Ducruet, P. De Langen and T. Notteboom. Aldershot: Ashgate, 175-89.

Frenken, K., van Oort, F. and Verburg, T. 2007. Related variety, unrelated variety and regional economic growth. *Regional Studies,* 41(5), 685-97.

Friedmann, J. 1966. *Regional Development Policy: A Case Study of Venezuela.* Cambridge: MIT Press.

Frost, J. 2008. *Gateways and Corridors: An Atlantic Perspective.* Ottawa: Association of Canadian Port Authorities.

Frost, J. and Roy, M.A. 2008. *Study on Potential Hub-and-Spoke Container Transshipment Operations in Eastern Canada for Marine movements of Freight (Short Sea Shipping) – Final Discussion Report.* Ottawa: CPCS Transcom Limited.

Fujita, M. and Mori, T. 1996. The role of ports in the making of major cities: Self-agglomeration and hub-effect. *Journal of Development Economics*, 49(1), 93-120.

Garver, M.S. and Mentzer, J.T. 1999. Logistics research methods: Employing structural equation modeling to test for construct validity. *Journal of Business Logistics*, 20, 33-57.

Gattorna, J. 2006. *Living Supply Chains*. New York, NY: Prentice Hall.

Gefen, D., Straub, D.W. and Bourdreau, M. 2000. SEM and regression: Guidelines for research practice. *Communication of the Association for Information Systems*, 4(7), 1-77.

Gereffi, G., Humphrey, J. and Sturgeon, T. 2005. The governance of global value chains. *Review of International Political Economy*, 12(1), 78-104.

Gillen, D., Parsons, G., Prentice, B. and Wallis, P. 2007. *Pacific Crossroads: Canada's Gateways and Corridors*. Paper to the Roundtable and International Conference, Asia-Pacific Gateway and Corridor Research Consortium: Regina, Canada, 21-22 Feb 2007. Available at: www.gateway-corridor.com [accessed: 22 May 2009].

Gillis, C. 2008. Breaking into the Jones Act: No easy feat, even with money and ambition to buy qualified vessels. *American Shipper*, 50(9), 71-3.

Giuliano, G. and O'Brien, T. 2008. Extended gate operations at the ports of Los Angeles and Long Beach: A preliminary assessment. *Maritime Policy and Management*, 35(2), 215-35.

Goetz, A.R. and Bandyopadhyay, S. 2007. *Regional Development Impacts on Trade Corridors: Recent Experiences from the United States*. Paper to the Asia-Pacific Gateway and Corridor Research Consortium: Regina, Canada, 21-22 Feb 2007. Available at: www.gateway-corridor.com [accessed: 22 May 2009].

Goldman Sachs. 2009. *Containerships: Big Bounce off the Bottom – What Next?* Singapore: Goldman Sachs Asia, 1 September 2009.

Gonzalez, M.M. and Trujillo, L. 2009. Efficiency measurement in the port industry: A survey of the empirical evidence. *Journal of Transport Economics and Policy*, 43(2), 157-91.

Gouvernal, E. 2002. L'organisation du transport maritime de ligne régulière et le rôle des ports. *Revue Transports*, 411, 15-29.

Gouvernal, E. and Daydou, J. 2005. Container rail freight services in North-west Europe: Diversity of organizational forms in a liberalizing environment. *Transport Reviews*, 25, 557-71.

Gouvernal, E. and Huchet, J.-P. 1998. La logistique des conteneurs: Le principal enjeu de l'industrie maritime de ligne régulière, in *ACTES La logistique, maîtrise du temps et de l'espace?*, edited by G. Fassio. Université de Nantes: IUT de Saint-Nazaire, 4ème trimestre, 77-87.

Gouvernal, E., Debrie, J. and Slack, B. 2005. Dynamics of change in the port system of the Western Mediterranean. *Maritime Policy and Management*, 32(2), 107-21.

Graf, A. and Maas, P. 2008. Customer value from a customer perspective: A comprehensive review. *Journal für Betriebswirtschaft*, 58(1), 1-20.

Grant, R.M. 2010. *Contemporary Strategy Analysis and Cases*. 7th Edition. Chichester: John Wiley and Sons.

Gross. I. 1997. Evolution in customer value, in *Service Quality: New Directions in Theory and Practice*, edited by R.T. Rust and R.L. Oliver. Thousand Oaks, CA: Sage.

Guardado, M., Frasquet, M. and Cervera, A. 2004. Benchmarking the port services: A customer oriented proposal. *Benchmarking: An International Journal*, 11(3), 320-30.

Gudeon, J. 2005. Consultant underscores need for Northeast highway and rail infrastructure. *Canadian Sailings*, October 24, 22-24.

Gunasekaran, A. and Kobu, B. 2007. Performance measures and metrics in logistics and supply chain management: A review of recent literature (1995-2004) for research and applications. *International Journal of Production Research*, 45(12), 2819-40.

Guy, E. and Lapointe, F. 2008. *Gateways and Corridor Initiatives: A Paradigm-Shift for Transportation Planning in Canada?* Proceedings of the 43rd Annual Conference of the Canadian Transportation Research Forum, 'Shaking up Canada's Transportation Systems to Meet Future Needs', Saskatoon, University of Saskatchewan Printing Services, 206-18.

Guy, E. and Lapointe, F. 2009. Politiques publiques pour le transport maritime sur le Saint-laurent: Cohérence des objectifs et mesures. Forthcoming report for Transport Quebec.

Guy, E. and Urli, B. 2007. *Structuration de problèmes complexes: Proposition méthodologique pour l'étude des politiques de transport maritime au Canada.* Présentation au: congrès annuel de l'Association des Sciences Administratives du Canada, Ottawa, 2-5 June 2007.

Guy, E. and Urli, B. 2008. *Décider d'un nouvel encadrement public pour le transport maritime: Structuration de problème dans les politiques publiques.* Conférence Outils pour Décider Ensemble 2008, Québec, 5-6 June 2008.

Guy, E. and Urli, B. 2009. *Analyse comparative des mesures d'intervention publique en support au transport maritime*. Québec: Transport Québec.

Hair, J.F., Black, W.C., Babin, B.J. and Anderson, R.E. 2010. *Multivariate Data Analysis: A Global Perspective*. New Jersey: Prentice Hall.

Hall, P.V. 2004. Mutual specialization, seaports and the geography of automobile imports. *Tijdschrift voor Economische en Sociale Geografie*, 95(2), 135-46.

Hall, P.V. 2007a. Seaports, urban sustainability and paradigm shift. *Journal of Urban Technology*, 14(2), 87-101.

Hall, P.V. 2007b. *Global Logistics and Local Dilemmas*. Paper to the International Conference on Gateways and Corridors, Vancouver, Canada, 3-4 May 2007. Available at: www.gateway-corridor.com [accessed: 22 May 2009].

Hall, P.V. 2009. Container ports, local benefits and transportation worker earnings. *Geojournal*, 74, 67-83.

Hamilton, C. 1999. *Measuring Container Port Productivity: The Australian Experience*. Background paper No. 17. Canberra: The Australia Institute, Australian National University.

Harrigan, K.R. 1984. Formulating vertical integration strategies. *Academy of Management Review*, 9(4), 638-52.

Hausman, A. 2001. Variations in relationship strength and its impact on performance and satisfaction in business relations. *Journal of Business and Industrial Marketing*, 16(7), 600-16.

Hayuth, Y. 1980. Inland container terminal: Functional and rationale. *Maritime Policy and Management*, 7, 283-9.

Hayuth, Y. 1981. Containerization and the load center concept. *Economic Geography*, 57, 160-75.

Hayuth, Y. 1982. Intermodal transportation and the hinterland concept. *Tijdschrift voor Economische en Sociale Geografie*, 73, 13-21.

Hayuth, Y. 1987. *Intermodality: Concept and Practice*. London: Lloyd's of London Press.

Hearn, G. and Pace, C. 2006. Value-creating ecologies: Understanding next generation business systems. *Foresight*, 8(1), 55-65.

Heaver, T.D. 1995. The implications of increased competition among ports for port policy and management. *Maritime Policy and Management*, 22(2), 125-33.

Heaver, T.D. 2002. The evolving roles of shipping lines in international logistics. *International Journal of Maritime Economics*, 4(3), 210-30.

Heaver, T.D. 2007. *Tying it all Together: The Future Logistics in and through Gateways*. Paper to the Vancouver International Conference, Asia-Pacific Gateway and Corridor Research Consortium, Vancouver, Canada, 2-3 May 2007. Available at: www.gateway-corridor.com [accessed: 31 May 2009].

Heaver, T.D., Meersman, H. and Van De Voorde, E. 2001. Cooperation and competition in international container transport: Strategies for ports. *Maritime Policy and Management*, 28(3), 293-305.

Henderson, J., Dicken, P., Hess, M., Coe, N. and Yeung, H.W-C. 2002. Global production networks and the analysis of economic development. *Review of International Political Economy*, 9(3), 436-64.

Hesse, M. 2010. Cities, material flows and the geography of spatial interaction: Urban places in the system of chains. *Global Networks*, 10(1), 75-91.

Hesse, M. and Rodrigue J.-P. 2004. The transport geography of logistics and freight distribution. *Journal of Transport Geography*, 12(3), 171-84.

Horbie, K. 2007. Corridors, gateways essential to compete in global economy, forum told. *Canadian Sailings*, July 9, 12-16.

Houston, G. 2008. *Vancouver Fraser Port Authority: The Making of an Amalgamated Port*. Ottawa: Port/Government Interface, Association of Canadian Port Authorities.

Hoyle, B.S. 1989. The port-city interface: Trends, problems and examples. *Geoforum*, 20(4), 429-43.

Hu, H.H., Kandampully, J. and Juwaheer, T.D. 2009. Relationships and impacts of service quality, perceived value, customer satisfaction and image: An empirical study. *The Services Industries Journal*, 29(2), 111-25.

Huemer, L. 2006. Supply management: Value creation, coordination and positioning in supply relationships. *Long Range Planning*, 39, 133-53.

Hull, A. 2005. Integrated transport planning in the UK: From concept to reality. *Journal of Transport Geography,* 13, 318-28.

HYW H2O. 2010. Highway H2O. Available at: http://www.hwyh2o.com [accessed: May 2009].

IBI Group. 2007. *Container Simulation Project for the BC Ministry of Transportation Container Trucking Forum*. Available at: http://www.th.gov.bc.ca/Container_Trucking/index.htm [accessed: May 2009].

IBI Group. 2008. *Étude sur le Corridor de Commerce Saint-Laurent—Grands Lacs. Développement Économique Canada.* Transport Québec et Société de Développement Économique du Saint-Laurent.

International Monetary Fund. 2009. World Economic Outlook.

Ircha, M.C. 2001. *US and Canadian Ports: Financial Comparisons*. Montreal: Canadian Transportation Act Review Committee.

Ircha, M.C. 2005. *Ithaque Terminal Sept-îles as a Proposed Container Transshipment Facility*. Sept îles Port Authority.

ISFORT. 1998. *Gioia Tauro: Logistica & Transhipment per lo sviluppo*. Roma: Gangemi Editore.

ISL (Institute of Shipping Economics and Logistics). 2006. *The Dry Bulk Market.* Bremen: ISL.

Jacobs, W. and Hall, P.V. 2007. What conditions supply chain strategies of ports? The case of Dubai. *GeoJournal*, 68(4), 327-42.

Jacobs, W., Ducruet, C. and De Langen, P. 2010. Integrating world cities into production networks: The case of port cities. *Global Networks*, 10(1), 92-113.

Jessop, B. and Sum, N-L. 2000. An entrepreneurial city in action: Hong Kong's emerging strategies in and for (inter-) urban competition. *Urban Studies*, 37(12), 2287-310.

Joan, J-M. 2008. Les enjeux liés à la mise en place des plates-formes intérieures pour l'acheminement terrestre des conteneurs en Europe du Nord-Ouest, in *Les Transports Maritimes dans la Mondialisation*, edited by J. Guillaume. Paris: L'Harmattan, 147-56.

Joignaux, G. and Courtois, A. 2009. La dimension territoriale d'un projet d'infrastructure fluviale: Le canal Seine-Nord Europe. *Les Cahiers Scientifiques du Transport*, 56, 87-107.

Kaplan, R.S. and Norton, D.P. 1996. *The Balanced Scorecard: Translating Strategy into Action*. Boston, MA: Harvard Business School Press.

Kaplinsky, R. and Morris, M. 2008. Value chain analysis: A tool for enhancing export supply policies. *International Journal of Technological Learning, Innovation and Development*, 1(3), 283-308.

Kavussanos, M. and Visvikis, I. 2006. *Derivatives and Risk Management in Shipping*. London: Whitherby Publishing.

Kennerley, M.P. and Neely, A.D. 2002. A framework of the factors affecting the evolution of performance measurement systems. *International Journal of Operations and Production Management*, 22(11), 1222-45.

Kent, J.L. and Flint, D.J. 1997. Perspectives on the evolution of logistics thought. *Journal of Business Logistics*, 18(2), 15-29.

Konings, R., Pielagen, B-J., Visser, J. and Wiegmans, B. 2009. *A New Hinterland Transport Concept for the Port of Rotterdam: Extended Gateways with Innovative Connecting Transport Systems*. Proceedings of the 88th Annual Transportation Research Board of the National Academies (TRB) Meeting (CDR), Washington, DC, paper 09-1786, 1-19.

Lakshmanan, T., Subramanian, U., Anderson, W. and Leautier, F. 2001. *Integration of Transport and Trade Facilitation*. Washington, DC: The World Bank.

Lam, J.S., Yap, W.Y. and Cullinane, K. 2005. Structure, conduct and performance on the major liner shipping routes. *Maritime Policy and Management*, 34(4), 359-81.

Lambert, D.M. and Cooper, M.C. 2000. Issues in supply chain management: Don't automate, obliterate. *Industrial Marketing Management*, 29(1), 65-83.

Lamming, R. 2002. *Sharing Secrets in Supplier Relationship Management*. Lecture for Silf Competence: Stockholm, 14 March 2002. Published in Utbildningsnytt, Silf Competence, May.

Langevin, A., Guy, E. and Lapointe, F. 2009. *Trafic conteneurisé et changements climatiques sur le Saint-Laurent: Évaluation du potentiel économique des ports du Saint-Laurent non soumis aux baisses de niveaux d'eau*. Unpublished research report for the Comité de concertation navigation durable.

Lavaud-Letilleul, V. 2005. L'aménagement de nouveaux terminaux à conteneurs et le renouvellement de la problématique flux-territoire dans les ports de la Rangée Nord. *Flux,* 59, 33-45.

Lawrence, D., Houghton, J. and George, A. 1997. International comparisons of Australia's infrastructure performance. *Journal of Productivity Analysis*, 8, 361-78.

Leach, P.T. 2008. No credit, no sales. *Journal of Commerce*, November 10.

Leitner, S.J. and Harrison, R. 2001. *The Identification and Classification of Inland Ports*. Report 0-4083-1. Austin, TX: Texas Department of Transportation.

Leone, L., Pergini, M., Bagozzi, R.P., Pierro, A. and Mannetti, L. 2001. Construct validity and generalizability of the Carver-White behavioural inhibition system/behavioural activation system scales. *European Journal of Personality*, 15, 373-90.

Leroux, G. 2008. *Shaking up Canada's Transportation System: Three Essential Priorities*. Fredericton: Canadian Transportation Research Forum.

Levinson, M. 2006. *The Box: How the Shipping Container Made the World Smaller and the World Economy Bigger*. Princeton, NJ: Princeton University Press.

Levinson, M. 2008. Freight Pain: The Rise and Fall of Globalization. *Foreign Affairs*, 87(6), 133-40.

Levy, D.T. 1985. The transaction cost approach to vertical integration: An empirical investigation. *Review of Economics and Statistics,* 67, 438-45.

Lim, S.-M. 1998. Economies of scale in container shipping. *Maritime Policy and Management*, 25, 361-73.

Lindgreen A. and Wynstra, F. 2005. Value in business markets: What do we know? Where are we going? *Industrial Marketing Management*, 34, 732-48.

Lirn, T.C., Thanopoulou, H.A. and Beresford, A.K.C. 2003. Transhipment port selection and decision-making behaviour: Analysing the Taiwanese case. *International Journal of Logistics: Research and Applications*, 6(4), 229-44.

Lo, F.C. and Salih, K. 1978. *Growth Pole Strategy and Regional Development Policy.* Oxford: Pergamon Press.

Macharis, C. 2002. Location analysis for Belgian intermodal terminal: A tool for achieving sustainable transport in Belgium, in *Current Issues in Port Logistics and Intermodality*, edited by T. Notteboom. Leuven: Garant, 99-123.

Magala, M. 2008. Modelling opportunity capture: A framework for port growth. *Maritime Policy and Management*, 35(3), 285-311.

Malaga, M. and Sammons, A. 2008. A new approach to port choice modelling. *Maritime Economics and Logistics*, 10(1-2), 9-34.

Mangan J., Proctor, A. and Gibbs, D. 2009. External influences on the Humber estuary ports, the largets concentration of port activity in the UK, in *Ports in Proximity – Competition and Coordination Among Adjacent Seaports*. Farnham: Ashgate, 225-33.

Mangan, J., Lalwani, C. and Fynes, B. 2007. *Roles for Sea Ports in the Context of Varying Supply Chain Strategies: The Development of Port Centric Logistics*. Proceedings of the 12th International Symposium of Logistics, Budapest, July 2007.

Markusen, A. 1996. Sticky places in slippery space: A typology of industrial districts. *Economic Geography*, 72(3), 291-314.

Marlow, P.B. and Paixão-Casaca, A.C. 2003. Measuring lean ports performance. *International Journal of Transport Management*, 1(4), 189-202.

Marsoft. 2009. *Containership Market Outlook*. September 2009. Singapore: Marsoft.

Martinez-Budria, A.E., Diaz-Armas, R., Nvarro-Ibanez, M. and Ravelo-Mesa, T. 1999. A study of the efficiency of Spanish port authorities using data envelopment analysis. *International Journal of Transport Economics*, 26 (2), 237-53.

Massey, D. 1979. In what sense a regional problem? *Regional Studies*, 13, 233-43.

McCalla, R.J. 1994. Canadian container ports: How have they fared? How will they do? *Maritime Policy and Management*, 21, 207-17.

McCalla, R.J. 1999a. From St. John's to Miami: Containerization at eastern seaboard ports. *Geoforum*, 48, 21-8.

McCalla, R.J. 1999b. Global change, local pain: Intermodal seaport terminals and their service areas. *Journal of Transport Geography*, 7, 247-54.

McCalla, R.J. 2008. Container transhipment at Kingston, Jamaica. *Journal of Transport Geography*, 16, 182-90.

McCalla, R.J., Slack, B. and Comtois, C. 2005. The Caribbean basin: Adjusting to global trends in containerization. *Maritime Policy and Management*, 32(3), 245-61

McMillan, C. 2006. *Embracing the Future: The Atlantic Gateway and Canada's Trade Corridor*. Vancouver: Asia Pacific Foundation of Canada.

Meidutė, I. 2005. Comparative analysis of the definition of logistics centres. *Transport*, 20, 106-10.

Memedovic, O., Ojala, L., Rodrigue, J.-P. and Naula, T. 2008. Fuelling the global value chains: What role for logistics capabilities? *International Journal of Technological Learning, Innovation and Development*, 1(3), 353-74.

Mentzer, J.T. and Konrad, B.P. 1991. An efficiency/effectiveness approach to logistics performance analysis. *Journal of Business Logistics*, 12(1), 33-62.

Mérenne-Shoumaker, B. 2007. La localisation des Grandes Zones Logistique. *Bulletin de la Société géographique de Liège*, 49, 31-40.

Morash, E.A. and Clinton, S.R. 1997. The role of transportation capabilities in international supply chain management. *Transportation Journal*, 36(3), 5-17.

Morgan, F. 1951. Observations on the study of hinterlands in Europe. *Tijdschrift voor Economische en Sociale Geografie*, 42, 366-75.

Muller, P. 2005. Esquisse d'une théorie du changement dans l'Action publique – structures, acteurs et cadres cognitifs. *Revue française de science politique*, 55(1), 115-87.

Muller, P. 2009. *Les Politiques Publiques*. 8th Edition. Paris: Presses Universitaires de France.

Naylor, B., Naim, M.M. and Berry, D. 1999. Leagility: Integrating the lean and agile manufacturing paradigms in the total supply chain. *International Journal of Production Economics*, 62(1/2), 107-18.

Neely, A. 1999. The performance measurement revolution: Why now and what next? *International Journal of Operations and Production Management*, 19(2), 205-28.

New Hampshire Business Review. 2008. *Hope Glimmers for Regional East-West Highway*. 15 February 2008.

Ng Koi Yu, A. 2006. Assessing the attractiveness of ports in the North European container transhipment market: An agenda for future research in port competition. *Journal of Maritime Economics and Logistics*, 8(3), 234-50.

Notteboom, T.E. 2000. *Spatial and Functional Integration of Container Port Systems and Hinterland Connections*. Round Table 113 of the European Conference of Ministers of Transport: Land Access to Seaports, Paris, 10-11 December 1998, 5-55.

Notteboom, T.E. 2002a. Consolidation and contestability in the European container handling industry. *Maritime Policy and Management*, 29, 257-69.

Notteboom, T.E. 2002b. Inland container shipping in North Europe: Spatial and organisational trends, in *Current Issues in Port Logistics and Intermodality*, edited by T. Notteboom. Leuven: Garant, 135-54.

Notteboom, T.E. 2004a. Container shipping and ports: An overview. *Review of Network Economics*, 3(2), 86-106.

Notteboom, T.E. 2004b. A carrier's perspective on container network configuration at sea and on land. *Journal of International Logistics and Trade*, 1(2), 65-87.

Notteboom, T.E. 2006a. The time factor in liner shipping services. *Maritime Economics and Logistics*, 8(1), 19-39.

Notteboom, T.E. 2006b. Traffic inequality in seaport systems revisited. *Journal of Transport Geography*, 14, 95-108.

Notteboom, T.E. 2006c. Port regionalization in Antwerp, in *Ports Are More Than Piers*, edited by T. Notteboom. Antwerp: De Lloyd, 309-27.

Notteboom, T.E. 2007. Concession agreements as port governance tools. *Research in Transportation Economics,* 17, 437-55.

Notteboom, T.E. 2008a. *The Relationship between Seaports and the Intermodal Hinterland in Light of Global Supply Chains: European Challenges*. Discussion Paper No. 2008-10, OECD – International Transport Forum, Paris, OECD.

Notteboom, T.E. 2008b. Bundling of freight flows and hinterland network development, in *The Future of Intermodal Freight Transport Operations Technology, Design and Implementation*, edited by R. Konings. Cheltenham: Elgar, 66-88.

Notteboom, T.E. 2009a. *Future Prospects of Container Shipping on the Asia-Europe Trade in Light of the Current Economic Crisis*. Proceedings of the 2009 International Conference on Shipping, Port, Logistics Management, Jungseok Research Institute, Inha University, Incheon, South Korea, 25-31.

Notteboom, T.E. 2009b. Path dependency and contingency in the development of multi-port gateway regions and multi-port hub regions, in *Ports in Proximity – Competition and Coordination among Adjacent Seaports*. Farnham: Ashgate, 55-72.

Notteboom, T.E. 2010. From multi-porting to a hub port configuration: The South-African container port system in transition. *International Journal of Shipping and Transport Logistics*, 2(2), 224-45.

Notteboom, T.E. and Rodrigue, J-P. 2004. *Inland Freight Distribution and the Sub-harborization of Port Terminals*. Proceedings of the First International Conference on Logistics Strategy for Ports, 22-26 September 2004. Edited by T. Notteboom. Dalian: Dalian Maritime University, 365-82.

Notteboom, T.E. and Rodrigue, J-P. 2005. Port regionalization: Towards a new phase in port development. *Maritime Policy and Management*, 32(3), 297-313.

Notteboom, T.E. and Rodrigue, J-P. 2008. Containerisation, box logistics and global supply chains: The integration of ports and liner shipping networks. *Maritime Economics and Logistics*, 10 (1/2), 86-106.

Notteboom, T.E., Ducruet, C. and De Langen, P.W. 2009. Introduction, in *Ports in Proximity – Competition and Coordination among Adjacent Seaports*. Farnham: Ashgate, 1-7.

Notteboom, T.E. and Rodrigue, J-P. 2009. The future of containerization: Perspectives from maritime and inland freight distribution. *GeoJournal*, 74, 7-22.

Notteboom, T.E. and Vernimmen, B. 2009. The effect of high fuel costs on liner service configuration in container shipping. *Journal of Transport Geography*, 17(5), 325-37.

Notteboom, T.E. and Winkelmans, W. 2001. Structural changes in logistics: How will port authorities face the challenge? *Maritime Policy and Management*, 28(1), 71-89.

O'Connor, K. 1989. Australian ports, metropolitan areas and trade-related services. *The Australian Geographer*, 20(2), 167-72.

O'Connor. K. 2008. Personal communication and unpublished manuscript.

O'Keefe, D. 2003.The future for Canada – US container port rivalries. *Statistics Canada*, Report 54F000XIE. Available at: http://dsp-psd.pwgsc.gc.ca/Collection/Statcan/54F0001X/54F0001XIE2003.pdf [accessed: March 2010].

O'Kelly, M.E. and Miller, H.J. 1994. The hub network design problem: A review and synthesis. *Journal of Transport Geography*, 2(1), 31-40.

OECD. 2008. *Port Competition and Hinterland Connections*. Paris: Joint Transport Research Centre of the OECD and the International Transport Forum, Discussion Paper No. 2008-19.

Olivier, D. 2005. Private entry and emerging partnerships in container terminal operations: Evidence from Asia. *Maritime Economics and Logistics*, 7, 87-115.

Olivier, D. and Slack, B. 2006. Rethinking the port. *Environmental Planning A*, 38(8), 1409-27.

Ontario-Quebec Continental Gateway and Trade Corridor. 2007. *Canada-Ontario-Quebec Memorandum of Understanding*. Available at: http://www.continentalgateway.ca/memorandum.html [accessed: 27 April 2009].

Pallis, A.A., Vitsounis, T.K. and De Langen, P.W. 2010. Port economics, policy and management: Review of an emerging research field. *Transport Reviews*, 30(1), 115-61.

Panayides, P.M. 2002. Economic organisation of intermodal transport. *Transport Reviews*, 22(4), 401-14.

Pando, J., Araujo, A. and Maqueda, F.J. 2005. Marketing management at the world's major ports. *Maritime Policy and Management*, 32(2), 67-87.

Pantouvakis, A., Chlomoudis, C. and Dimas, A. 2008. Testing the SERVQUAL scale in the passenger port industry: A confirmatory study. *Maritime Policy and Management*, 35(5), 449-67.

Park, R.K. and De, P. 2004. An alternative approach to efficiency measurement of seaports. *Journal of Maritime Economics and Logistics*, 6(1), 53-69.

Parola, F., Lee, S-W. and Ferrari, C. 2006. On the integration of logistics activities by shipping lines: The case of East Asia. *Journal of International Logistics and Trade*, 4(1), 109-30.

Parr, J. 1999. Growth-pole strategies in regional economic planning: A retrospective view. *Urban Studies*, 36(7), 1195-215.

Patton, D. General cargo hinterlands of the ports of New York, Philadelphia, Baltimore and New Orleans. *Annals of the Association of American Geographers*, 48, 436-55.

Penty, R. 2009. Gateway council created: The region's prominent business leaders are teaming up for advisory committee. *Saint John Telegraph Journal*, 6 February.

Perroux, F. 1950. Economic space: Theory and applications. *Quarterly Journal of Economics*, 64, 89-104.

Peters, T. 2007. New container port in the works for Nova Scotia. *Canadian Sailings*, June 18, 14-15.

Peters, T. 2008a. Container terminal in the works for Sydney. *Canadian Sailings*, 2 June 2008.

Peters, T. 2008b. Support grows for Atlantic Canada gateway advisory council. *Canadian Sailings*, December 15, 29.

Peters, T. 2009a. Melford container terminal to open in 2012. *Halifax Chronicle Herald*, 2 September.

Peters, T. 2009b. Halifax port, seaway talk short-sea shipping. *Canadian Sailings*, March 9, 13.

Pettit, S.J. 2008. United Kingdom ports policy: Changing government attitudes. *Marine Policy*, 32(4), 719-27.

Pettit, S.J. and Beresford, A.K.C. 2009. Port development: From gateways to logistics hubs. *Maritime Policy and Management*, 36(3), 250-65.

Port of Seattle. 2009. *Ports of Seattle, Tacoma Strategize for Growth, Two ports Tackle Challenges, Opportunities in Changing Competitive Landscape*. Press Release, 12 October 2009.

Porter, M.E. 1985. *Competitive Advantage: Creating and Sustaining Superior Performance*. New York, NY: The Free Press.

Potter, S. and Skinner, M. 2000. On transport integration: A contribution to better understanding. *Futures*, 32, 275-87.

PRI. 2008. *The Emergence of Cross-Border Regions Between Canada and the United States – Final Report*. Ottawa: Policy Research Initiative.

Rabinovich, E., Windle, R., Dresner, M. and Corsi, T. 1999. Outsourcing of integrated logistics functions: An examination of industry practices. *International Journal of Physical Distribution and Logistics Management*, 29(6), 353-74.

Randers, J. and Göluke, J. 2007. Forecasting turning points in shipping freight rates: Lessons from 30 years of practical effort. *System Dynamics Review*, 23 (2/3), 253-84.

Ridolfi, G. 1986. Réseaux et nœuds pour le transport multimodal et combiné en Italie, in *Ports et Mers*, edited by J. Charlier. Caen: Paradigme, 285-96.

Rimmer, P.J. 1977. A conceptual framework for examining urban and regional transport needs in Southeast Asia. *Pacific Viewpoint*, 18, 133-47.

Rizet, C. 1997. Coûts portuaires: Contraintes et organisation, in *Politiques de Transport et Compétitivité*, edited by E. Gouvernal, M. Guilbault and C. Rizet. Paris: Hermès, 89-105.

Robinson, R. 2002. Ports as elements in value-driven chain systems: The new paradigm. *Maritime Policy and Management*, 29(3), 241-55.

Rodrigue, J-P. 2007. *Gateways, Corridors, and Global Freight Distribution: The Pacific and the North American Maritime/Land Interface*. Paper to the Vancouver International Conference, Asia-Pacific Gateway and Corridor Research Consortium, Vancouver, Canada, 2-3 May 2007. Available at: www.gateway-corridor.com [accessed: 22 May 2009].

Rodrigue, J-P. and Browne, M. 2008. International maritime freight movements, in *Transport Geographies: Mobilities, Flows and Spaces*, edited by R. Knowles et al. Oxford: Blackwell, 156-78.

Rodrigue, J-P. and Notteboom, T. 2007. Re-assessing port-hinterland relationships in the context of global supply chains, in *Ports, Cities, and Global Supply Chains*, edited by J. Wang, D. Olivier, T. Notteboom and B. Slack. Aldershot: Ashgate, 51-66.

Rodrigue, J-P. and Notteboom, T. 2009a. The terminalization of supply chains: Reassessing port-hinterland logistical relationships. *Maritime Policy and Management*, 36(2), 165-83.

Rodrigue, J-P and Notteboom, T. (eds) 2009b. Containerization in a globalized world (special issue). *GeoJournal*, 74, 1-83.

Rodrigue, J-P and Notteboom, T. 2009c. Transport terminals, in *The Geography of Transport Systems*, 2nd Edition, edited by J-P. Rodrigue, C. Comtois and B. Slack. London: Routledge, 126-43.

Roll, Y. and Hayuth, Y. 1993. Port performance comparison applying data envelopment analysis (DEA). *Maritime Policy and Management*, 20(2), 153-61.

Roso, V., Woxenius, J. and Lumsden, K. 2009. The dry port concept: Connecting container seaports with the hinterland. *Journal of Transport Geography*, 17, 338-45.

Rossall, E. and Barrie, D. 2008. *European Distribution Report 2008*. London: Cushman and Wakefield.

Roy, M.A. and Frost, J. 2009. Unprecedented opportunities to promote short-sea shipping in Canada. *Canadian Sailings*, February 9, 15-16.

Ryoo, D.K. and Thanopoulou, H.A. 1999. Liner alliances in the globalisation era: A strategic tool for Asian container carriers. *Maritime Policy and Management*, 26, 349.

Sanchez, R.J., Hoffman, J., Micco, A., Pizzolitto, G., Sgut, M. and Wilmsmeier, G. 2003. Port efficiency and international trade: Port efficiency as a determinant of maritime transport costs. *Journal of Maritime Economics and Logistics*, 5(2), 199-218.

Savy, M. 2005. Les plates-formes logistiques. *Logistiques Magazines*, Numéro spécial *20 ans de logistique*, 201, 110-14.

Senate of Canada. 2008. *Time for a New National Vision – Opportunities and Constraints for Canada in the Global Movement of Goods*. Ottawa: Senate Standing Committee on Transport and Communications. Available at: http://www.parl.gc.ca/39/2/parlbus/commbus/senate/Com-e/TRAN-E/rep-e/rep07jun08-e.htm [accessed: 20 April 2009].

Short, J., Breitbach, C.and Essex, J. 2000. From world cities to gateway cities. *City*, 4(3), 317-40.

Slack, B. 1985. Containerization, inter-port competition, and port selection. *Maritime Policy and Management*, 12(4), 293-303.

Slack, B. 1989a. Port services, ports and the urban hierarchy. *Tijdschrift voor Economische en Sociale Geografie*, 80(4), 236-43.

Slack, B. 1989b. The port service industry in an environment of change. *Geoforum*, 20(4), 447-57.

Slack, B. 1989c. Gateway or cul-de-sac? The saint lawrence river and eastern Canadian container traffic. *Etudes Canadiennes*, 26, 49-58.

Slack, B. 1990. Intermodal transportation in North America and the development of inland load centers. *Professional Geographer*, 12, 307-10.

Slack, B. 1993. Pawns in the game: Ports in a global transportation system. *Growth and Change*, 24, 579-88.

Slack, B. 1994. Domestic containerisation and the load centre concept. *Maritime Policy and Management*, 21, 229-36.

Slack, B. 1998. Intermodal transportation, in *Modern Transport Geography*, 2nd Edition, edited by B. Hoyle and R. Knowles. Chichester: John Wiley & Sons, 263-90.

Slack, B. 1999. Satellite terminals: A local solution to hub congestion? *Journal of Transport Geography*, 7, 241-6.

Slack, B. 2007. The terminalisation of seaports, in *Ports, Cities and Global Supply Chains*, edited by J. Wang, D. Olivier, T. Notteboom and B. Slack. Aldershot: Ashgate, 41-50.

Slack, B. and Frémont, A. 2009. Fifty years of organisational change in container shipping: Regional shift and the role of family firms. *Geojournal*, 74(1), 23-34.

Slack, B. and Wang, J.J. 2002. The challenge of peripheral ports: An Asian perspective. *GeoJournal*, 56, 159-66.

Slack, B., Comtois, C. and Sletmo, G. 1996. Shipping lines as agents of change in the port industry. *Maritime Policy and Management*, 23, 289-300.

Sletmo, G.K. 1999. Port life cycle: Policy and strategy in the global economy. *International Journal of Maritime Economics*, 1, 11-37.

Song, D-W. and Panayides, P.M. 2008. Global supply chains and port/terminal: Integration and competitiveness. *Maritime Policy and Management*, 35(1) 73-87.

Soppé, M., Parola, F. and Frémont, A. 2009. Emerging inter-industry partnerships between shipping lines and stevedores: From rivalry to cooperation? *Journal of Transport Geography*, 17(1), 10-20.

Spekman, R.E. and Davis, E.W. 2004. Risky business: Expanding the discussion on risk and the extended enterprise. *International Journal of Physical Distribution and Logistics Management*, 34(5), 414-33.

Spiteri, J.M. and Dion, P.A. 2004. Customer value, overall satisfaction, end-user loyalty and market performance in detail intensive industries. *Industrial Marketing Management*, 33, 675-87.

State of Victoria, 2009. *Toward an Integrated and Sustainable Transport Future: A New Legislative Framework for Transport in Victoria*. Melbourne: State of Victoria. Available at: http://www.transport.vic.gov.au/DOI/DOIElect.nsf/$UNIDS+for+Web+Display/336EB666427AE28FCA25760200221400/$FILE/Towards%20an%20integrated%20and%20sustainable%20transport%20future%20WEB.pdf [accessed: 14 January 2010].

Stopford, M. 2009. *Maritime Economics*. 3rd Edition. London and New York, NY: Routledge.

Stuckey, J. and White, D. 1993. When and when not to vertically integrate. *Sloan Management Review*, 34(3), 71-83.

Su, Y., Liang, G.S., Liu, C.F. and Chou, T.Y. 2003. A study on integrated port performance comparison based on the concept of balanced scorecard. *Journal of Eastern Asia Society for Transportation studies*, 5, 609-24.

Surel, Y. 2000. The role of cognitive and normative frames in policy-making. *Journal of European Public Policy*, 7(4), 495-512.

Swift, A. 2009. CEO prepares port for future economic recovery. *Canadian Sailings*, January 26, 8-9.

Taaffe, E.J., Morrill, R.L. and Gould, P.R. 1963. Transport expansion in underdeveloped countries: A comparative analysis. *Geographical Review*, 53(4), 503-29.

Talley, W.K. 1994. Performance indicators and port performance evaluation. *Logistics and Transportation Review*, 30, 339-52.

Taylor, P.J. 2001. Specification of the world city network. *Geographical Analysis*, 33(2), 181-94.

Taylor, P.J., Catalano, G. and Walker, D.R.F. 2002. Measurement of the world city network. *Urban Studies,* 39, 2367-76.

Taylor, R.B. and Hackston, D.C. 2008. Southern ontario gateway transportation and logistics issues.

Tikkanen, H. and Alajoutsijarvi, K. 2002. Customer satisfaction in industrial markets: Opening up the concept. *Journal of Business and Industrial Marketing*, 17(1), 22-45.

Tongzon, J.L. 2001. Efficiency measurement of selected Australian and other international ports using data envelopment analysis. *Transportation Research Part A*, 35, 107-22.

Tongzon, J.L. 2008. Port choice and freight forwarders. *Transportation Research Part E*, 45(1), 186-95.

Tongzon, J.L. and Ganesalingam, S. 1994. An evaluation of ASEAN port performance and efficiency. *Asian Economic Journal*, 8, 317-30.

Tongzon, J.L. and Heng, W. 2005. Port privatization, efficiency and competitiveness: Some empirical evidence from container ports (terminals). *Transportation Research Part A*, 39(5), 405-24.

Tovar, B., Jara-Diaz, S. and Trujillo, L. 2007. Econometric estimation of scale and scope economies within the port sector: A review. *Maritime Policy and Management*, 34(3), 203-23.

Transport Canada. 2007. *National Policy Framework for Strategic Gateways and Trade Corridors*. Ottawa: Transport Canada.

Transport Canada. 2009a. *Transportation in Canada: An Overview*. Ottawa: Transport Canada. Available at: http://www.tc.gc.ca/policy/report/aca/anre2008/pdf/addendum.pdf [accessed: 5 March 2010].

Transport Canada. 2009b. *Canada's Gateways*. Available at: http://www.canadasgateways.gc.ca/index2.html [accessed: 12 May 2009].

Tuppen, J. 1983. *The Economic Geography of France*. London: Croom Helm.

Ugboma, C., Ogwude, I.C., Ugboma, O. and Nnadi, K. 2007. Service quality and satisfaction measurements in Nigerian ports: An exploration. *Maritime Policy and Management*, 34(4), 331-46.

UNCTAD (United Nations Conference on Trade and Development). 1982. *Improving Port Performance: Management of General Cargo Operations* (trainee's workbook). Cardiff: Drake Educational Associates Ltd.

UNCTAD (United Nations Conference on Trade and Development). 1983a. *Development and Improvement of Ports: Development of Bulk Terminals*. Geneva: UNCTAD.

UNCTAD (United Nations Conference on Trade and Development). 1983b. *Manual on a Uniform System of Port Statistics and Performance Indicators*. Geneva: UNCTAD.

UNCTAD (United Nations Conference on Trade and Development). 2006. *Trade Facilitation Handbook, Part 1, National Facilitation Bodies: Lessons from Experiences.* Geneva: UNCTAD.

UNCTAD (United Nations Conference on Trade and Development). 2007. *Review of Maritime Transport 2007.* Geneva: UNCTAD.

UNCTAD (United Nations Conference on Trade and Development). 2009. *Review of Maritime Transport 2009*. Geneva: UNCTAD.

USDOT-MARAD (United States Department of Transportation, Marine Administration). 2000. *Marine Terminal Productivity Measures*. Reston: Transystems Corporation.

Van Breedam, A. and Vannieuwenhuyse, B. 2006. The extended gateway: A new project for logistics flanders, in *Ports Are more than Piers*, edited by T. Notteboom. Antwerp: De Lloyd, 289-305.

Van der Horst, M.R. and De Langen, P.W. 2008. Coordination in hinterland transport chains: A major challenge for the seaport community. *Maritime Economics and Logistics*, 10, 108-29.

Van der Lugt, L. and De Langen, P. 2005. The changing role of ports as locations for logistic activities. *Journal of International Logistics and Trade*, 3(2), 59-72.

Van der Lugt, L., De Langen, P. and Hagdorn, L. 2009. *Value Creation and Value Capture in the Ports Business Ecosystem*. International Association of Maritime Economists (IAME) Conference, (proceedings: CD-Rom), Copenhagen, 24-26 June 2009.

Van Klink, H.A. 1995. *Towards the Borderless Mainport Rotterdam: An analysis of Functional, Spatial and Administrative Dynamics in Port Systems*. Amsterdam: Tinbergen Institute, Research Series.

Van Klink, A. and Van der Berg, G.C. 1998. Gateways and intermodalism. *Journal of Transport Geography*, 6, 1-19.

Van Pelt, M. 2003. *Moving Trade: An Introduction to Trade Corridors*. Hamilton: Work Research Foundation.

Vernon, R.1966. International investment and international trade in the product cycle. *Quarterly Journal of Economics*, 80(2), 190-207.

Vickerman, R. 2007. *Gateways, Corridors and Competitiveness: An Evaluation of Trans-European Networks and Lessons for Canada*. Paper to the Winnipeg Roundtable Conference, Winnipeg, Canada, 21-22 Feb 2007 for Asia-Pacific Gateway and Corridor Research Consortium, Canada. Available at: www.gateway-corridor.com [accessed: 22 May 2009].

Vigarié, A. 1999. From breakbulk to containers: The transformation of general cargo handling and trade. *GeoJournal*, 48, 3-7.

Vigarié, A. 2004. L'évolution de la notion d'arrière-pays en économie portuaire. *Transports*, 428, 372-87.

Vogt, J.J., Pienaar, W.J. and de Wit, P.W.C. 2005. *Business Logistics Management*. Oxford: Oxford University Press.

Walters, D. 2004. New economy – new business models – new approaches. *International Journal of Physical Distribution and Logistics Management*, 34(3/4), 219-29.

Wang, J. 2009. Hong Kong in transition from a hub port city to a global supply chain management, in *Ports in Proximity: Competition and Coordination Among Adjacent Seaports*, edited by C. Ducruet, P. De Langen and T. Notteboom. Aldershot: Ashgate, 261-72.

Wang, J., Olivier, D., Notteboom, T. E. and Slack, B. (eds) 2007. *Ports, Cities and Global Supply Chains*. Aldershot: Ashgate.

Wang, T-F. and Cullinane, K. 2006. The efficiency of European container terminals and implications for supply chain management. *Maritime Economics and Logistics*, 8, 82-99.

Weibe, J.D. 1993. *Asia Pacific Transport: Blueprint for a Canadian Gateway*. Issues 2, Summer. Vancouver: Asia Pacific Foundation of Canada.

Weigand, G. 1958. Some elements in the study of port geography. *Geographical Review*, 48, 185-200.

Wheeler, D. and McPhee, W. 2006. Making the case for the added-value chain. *Strategy and Leadership*, 34(4), 39-46.

Williams, C. 2007. Proposed container terminal in Maine raises eyebrows. *Canadian Sailings*, December 24, 25.

Wilson, G.N. and Summerville, J. 2008. Transformation, transportation or speculation? The Prince Rupert container port and its impact on Northern British Columbia. *Canadian Political Science Review*, 2(4), 1-13.

Wolfe, D. and Gertler, M. 2004. Clusters from the inside and out: Local dynamics and global linkages. *Urban Studies*, 41(5-6), 1071-93.

Woo, S-H. and Pettit, S.J. 2009. Port performance in changing logistics environment, in *The International Handbook of Maritime Economics*. London: Edward Elgar (in press).

Woo, S-H., Pettit, S.J and Beresford, A.K.C. 2008. *A New Port Performance Measurement Framework in a Changing Logistics Environment*. Proceedings of the LRN 2008 Annual Conference, Liverpool, UK, 141-6.

Woodruff, R.B. 1997. Customer value: The next source for competitive advantage. *Journal of the Academy of Marketing Science*, 25(2), 139-53.

Woodruff, R.B. and Flint, D.J. 2003. Research on business-to-business customer value and satisfaction. *Evaluating Marketing Actions and Outcomes Advances in Business Marketing and Purchasing*, 12, 515-47.

Zohil, J. and Prijon, M. 1999. The MED rule: The interdependence of container throughput and transhipment volumes in the Mediterranean ports. *Maritime Policy and Management*, 26(2), 175-93.

Index

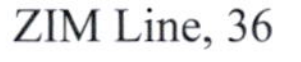